DATE DUE

Coal Firms under the New Social Regulation

Richard A. Harris   *Duke Press Policy Studies*

Duke University Press   Durham 1985

© 1985 Duke University Press, all rights reserved
Printed in the United States of America on acid-free paper
Library of Congress Cataloging-in-Publication Data
Harris, Richard A., 1951–
Coal firms under the new social regulation.
(Duke Press policy studies)
Bibliography: p.
Includes index.
1. Coal trade—government policy—United States.
2. Coal mines and mining—Safety regulations—United
States. 3. Coal—Environmental aspects—Government
policy—United States. I. Title. II. Series.
HD9546.H36 1985   338.7′622334′0973   85-13056
ISBN 0-8223-0648-4

To Ellen Carol Wolf, whose strength,
encouragement, insight,
and love have inspired the same in me.

Contents

Preface

Americans have always recognized the legitimacy of some degree of central government intervention in commercial affairs. Indeed, the divisiveness of state commercial policies under the Articles of Confederation was the proximate cause of the founding fathers' deliberations first in Annapolis during 1786 and subsequently in Philadelphia during 1787. While the founders intended to rely on free enterprise to provide a satisfactory level of goods and services, they also saw a need for the government to play a positive role in commercial affairs in order to provide for the general welfare. In the course of the two centuries since the framing of the Constitution this positive role has expanded from a baseline of monitoring interstate commerce to encompass an array of functions and responsibilities clearly not contemplated by the founders.

As federal regulatory authority has expanded to include control over price and entry into markets, labor practices, product quality and eventually, environmental impacts, political scientists have devoted increasing attention to the politics of regulatory policy. These efforts, while yielding critical insights, have been dominated by two important perspectives, both of which have significant theoretical implications. First, these studies have concentrated on the political influence wielded by business. Though an entirely reasonable research focus, it can lead to an exaggeration of business' influence, a minimization of the role played by other actors in regulatory politics and an underestimation of the possibility for major shifts in regulatory policy. Second, these studies almost universally treat entire industries or even business in general as unitary actors. Such a methodological assumption also tends to support the notion of strong business influence in regulatory politics. Finally, as notions of iron triangles, agency capture and, most significantly, interest group liberalism gained credence, political scientists and their audiences have come to take for granted an inherent conservatism and unresponsiveness in regulatory affairs. The New Deal is seen as the last major response to demands for regulating business—except, of course, for those demands emanating from business itself.

This study departs from these traditional perspectives in arguing that major changes have been brought about in federal regulatory policy during the past ten to fifteen years and in maintaining that individual firms rather than a larger unit of analysis (e.g., an industry) is the best focus for an investigation of regulatory politics. Further, I attempt to show that another major shift in regulatory politics and policy occurred in the late 1960s and early 1970s: the emergence of so-called "social regulation."

In analyzing the politics of regulatory policy under this new regulation, I treat individual firms as the primary business participants. This focus on the firm derives, in part, from the nature of new regulation which is organized functionally, imposing uniform rules that cut across firms in systematic ways. It is wrongheaded, I argue, to impute a community of interest, let alone unity of action, to collectivities of firms that are different in size, location and makeup. At a more profound level, though, focusing on the firm is a response to serious criticism by economic theorists of traditional political science studies of regulation. Mancur Olson, in particular, has made a persuasive argument against group-based political theory, an argument that this study will develop and adapt to a specific substantive case. In addition, economic theories of regulation based on microeconomic models of behavior suggest the firm rather than a larger unit of analysis as the appropriate object of inquiry.

This is not to suggest that we drop political theory and adopt economic theory. The latter has weaknesses too. In particular, its gains in empirical theory are grounded on assumptions that minimize the complexity and the capacity for macro change in the political system. While the bias against major regulatory change afflicts both economic and political theory—though for different reasons—political science has much to say about the complexity and structure of regulatory politics. Accordingly, this study develops a framework of analysis that draws on both economic and political theory. The result of this synthesis is an investigation of firm preferences and behaviors in the political context of the new social regulation.

The investigation is developed by examining the enactment and implementation of the Surface Mining Control and Reclamation Act of 1977 (SMRCA). While this case study approach restricts the power and reliability of my conclusions about regulatory politics, it does clarify behavioral patterns of firms and other regulatory actors. It also generates strong hypotheses for future, more comprehensive investigations. Moreover, our relatively short experience with regulation such as SMCRA warrants a case study approach since research is still coming to grips with the character of this kind of regulation.

On a final note about the genesis of this research, it should be pointed out that it was completed in early 1981, before Ronald Reagan had a chance to put his imprimatur on federal regulatory policy. This is significant since a major conclusion of the study is that the new social regulation was consciously designed to guard against incursions from hostile chief executives as well as capture by businessmen. In this regard, the Reagan policy on regulation provided a convenient test of the new regime's staying power, at least with respect to SMCRA. Indeed, the Reagan appointee to head the Office of Surface Mining, the agency responsible for SMCRA, was an adamant opponent of the legislation. Consequently an epilogue has been added which attempts to document firm behavior and the fate of SMCRA in this hostile environment. This assessment is necessarily more speculative.

I, of course, would be remiss in not acknowledging those without whose support this project could not have been brought to fruition. Foremost among those to whom I am indebted is my dissertation advisor at the University of Pennsylvania, Jack H. Nagel. His attention to detail and insistence on excellence were matched only by his unflagging support and encouragement. Any professional contribution made by this research is largely a result of his guidance and counsel. In addition, I must thank Edwin Haefele and William Kristol for their comments and criticism. Frederick Thompson of Columbia also offered helpful advice as well as an opportunity to present this work in condensed form at the Association for Public Policy Analysis and Management meetings in 1982. I am especially grateful to Elizabeth Dean and Rose Kenerley for their patience and professionalism in typing the various drafts of the manuscript. Finally, this project could not have been successfully completed without the generosity of the Department of Political Science at Penn, which underwrote the cost of my not insignificant phone bills for cross-country interviews, and the Brookings Institution, which supported me under their Visiting Scholars Program. The use of an office, telephone and Brookings letterhead actually were critical in gaining access to officials and institutions. Moreover, the Brookings cafeteria proved to be the most congenial and effective venue for carrying out numerous interviews in Washington. While I gratefully acknowledge all of the assistance and support of those mentioned, I of course take full responsibility for the contents of this manuscript.

Richard A. Harris

1 Introduction

In August 1977 President Jimmy Carter signed into law the Surface Mining Control and Reclamation Act (SMCRA). This law established a regulatory program covering the surface mining of coal nationwide and funded study of the prospects for extending the program to regulate the surface mining of hard-rock minerals such as copper, iron and gravel. In light of SMCRA's legislative history, the media, many legislators and supporters of the bill portrayed this as a victory for environmentalists and a defeat for the coal industry. To some extent this was an accurate perception. The struggle to enact SMCRA had dragged on for nearly a decade, the legislation surviving some bitter congressional confrontations between environmentalists and coal operators, two vetoes by President Gerald Ford and, ultimately, a Supreme Court challenge. Such a political contest tempts one to speak of the public interest, namely environmental quality, triumphing over a private interest, coal mining.

This view of SMCRA's enactment, though, represents at best a gross oversimplification and at worst a fundamental misperception of regulatory politics. On the one hand, the equation of environmental laws with the public interest is open to challenge. The "public interest" is a slippery term; environmental issues ordinarily raise competing public interests. On the other hand—and this is the critical point of this research—students of regulatory politics must be equally cautious about viewing an industry or some larger aggregation of business firms as a unified political actor. As Elmer Smead notes in a study of business regulation, ". . . government regulation may have contradictory effects; what hurts one business may help another" (Smead, 1969, p. 142). This being the case, it makes sense to treat individual firms rather than larger units of analysis as business participants in regulatory politics.

If a firm-level strategy of inquiry seems advisable for the study of regulatory politics generally, it is imperative in dealing with the so-called "new social regulation." This newer regulation that emerged in the late 1960s and early 1970s represents not simply more federal regulation, but a qualitative shift that has led the central government to regulate in new issue areas (Lilley and Miller, 1977). Regulation has become a legitimate policy instrument in the pursuit of social goals such as environmental quality as well as traditional economic goals such as competition and fair prices.

The new social regulation has resulted, according to some observers, in major changes in government-business relations (Haefele, 1978). Perhaps most important, this regulation is organized along functional rather than industry-specific lines. It seeks to regulate all business activity bearing on environmental quality, consumer affairs, economic opportunity and the like. A logical outcome of this functional organization is the imposition of uniform performance standards that cut across individual firms as well as entire industries.

Given this functional rather than industry-specific structure, analysis of regulatory politics demands a firm-level approach. Viewing business aggregates as unified actors may be a legacy of earlier regulatory institutions designed to control price and entry in particular industries. Common sense, however, suggests that individual firms as rational decision makers seeking profit maximization will view uniform performance standards differently depending on the particular political, economic, and financial circumstances confronting management. While a number of other characteristics distinguish the new social regulation from earlier forms, its functional organization and attendant uniform performance standards constitute the most significant change with respect to understanding regulatory politics and policy. The combination of uniform standards and interfirm differences leads to exactly what Smead suggested, contradictory regulatory impacts at the firm level. Therefore, it is especially inappropriate to approach the study of regulatory politics in this new setting from something other than a firm-level perspective.

SMCRA typifies, indeed epitomizes in many respects, the new regulatory politics and policy that have emerged in the last decade-and-a-half. Like other examples of the new social regulation it cut across firms in markedly different circumstances and imposes uniform performance standards on those firms. Moreover, SMCRA reflects the socioeconomic trade-offs, underlying ideological perspective, political strategies, and policy prescriptions characteristic of the new social regulation.

The Case of SMCRA

During the 1974 Senate hearings on SMCRA (S.425), a number of individuals including legislators referred to the United States as "the Saudi Arabia of coal." Highly suggestive and, no doubt, calculated to appeal to potential anti-SMCRA forces in Congress, this was an apt analogy. Approximately 32 percent of the world's proven coal reserves lie within America's borders (House Hearings, 1972a). Coal accounts for nearly three-fourths of our fossil fuel resources (Senate Hearings, 1972b). At 1979 production levels, U.S. proven reserves would last just over two hundred years (President's Commission on Coal, 1980).

These facts, when considered in light of America's energy problems since the oil embargo of 1973, naturally focused the attention of national policymakers on the coal industry. A revitalized coal industry could provide a "bridge" to renewable energy resources in the twenty-first century. The emphasis accorded coal production by the Ford and Carter administrations reflected this focus. Congress, in addition, mandated industrial and utility conversion to coal power in the Power Plant and Industrial Fuel Use Act of 1978 (P.L. 95-620). Furthermore, Congress directed DOE to implement this law with dispatch under the Energy Supply and Environmental Coordination Act of 1974 (P.L. 93-319). Even if, for political purposes, coal's importance had been exaggerated as something of a panacea or trump card to be played against OPEC, it would have a great deal to do with America's medium-run energy future.

However pressing the perceived need to expand coal output, it constituted only one of two components in the political climate surrounding SMCRA. Obviously, environmentalists, a potent political force by the early 1970s, could not agree to a national policy of expansion at any cost. From a political standpoint this renewed attention to coal created an interesting situation because it coincided with the triumph of environmentalism. By the mid-1970s the environmental movement had brought to fruition several major policy initiatives and established itself in the federal bureaucracy. The existence of the National Environmental Protection Act (NEPA), 1972 Water Pollution Control Act Amendments (WPCAA), Clean Air Act Amendments Acts of 1970 and 1977 (P.L. 91-604 & 95-95), Council on Environmental Quality (CEQ) and the Environmental Protection Agency (EPA) all indicated that the issue of environmental protection had been placed squarely on the national agenda for the foreseeable future. SMCRA was another of environmentalism's attempts to control, through federal regulation, business behavior that produced ecological externalities.

Efforts to regulate surface mining clashed not only with the interests of mining firms, especially coal, but also with attempts of national leaders to promote a rapid expansion of coal production after 1973. Advocates of expansion saw surface mining as a safe and highly productive alternative to deep mining. By conservative estimates, surface mining can recover 80 percent of the coal in a deposit as compared to 57 percent for deep mining (Report to Senate Committee on Interior and Insular Affairs, 1974). Because of surface mining's high productivity and recovery rate and because practically all of the rich western deposits were amenable to surface techniques, in the eyes of many, SMCRA posed a direct threat to plans for increased coal production, not to mention the political objective of energy independence. Naturally, these fears were magnified when early versions of SMCRA raised the prospect of a total ban on surface mining.

Understandably, national commitments to rapid coal mining development and environmental protection gave rise to a certain policy ambivalence. The statement of findings and purposes in SMCRA makes clear the fact that policymakers sought to reconcile the goals of accelerated coal production and a clean environment: "It is the purpose of this Act to . . . assure that coal supplies essential to the Nation's energy requirements, and to its social and economic well-being are provided and to strike a balance between protection of the environment . . . and the Nation's need for coal as an essential source of energy" (30 USC, sec. 1202). Echoing this approach, Secretary of Interior Cecil Andrus asserted in 1977: "Increasing the Nation's ability to produce and use coal in order to decrease our reliance on imported oil and scarce natural gas is essential. With sound environmental safeguards, surface mining will be an acceptable way to produce much of the coal that will be needed to meet this demand" (House Hearings, 1977b). The seven years it took to enact SMCRA and the continuing administrative and legal battles over its implementation bear witness to the difficulty of translating the principle of reconciliation into practice.

The concept of a federal regulatory program for surface mining first entered the national agenda in 1968. Not until 1977, however, did Congress successfully enact SMCRA. Between 1971 and 1977, more than twenty-five separate bills were introduced; six made it out of committee and passed at least one house of Congress; and two were vetoed. This tortuous legislative history spanned the crucial years in which the new regulatory regime took shape. Significantly, it also coincided with the energy crisis and declining productivity in the U.S. economy. These latter two phenomena posed an extraordinary challenge to environmentalism insofar as the new regulation threatened to exacerbate these problems. Therefore, the legislative history of SMCRA offers a good opportunity to examine the new social regulation.

The critical issues, the political strategies and the key participants in the emergence of the new social regulation all appear in SMCRA's genesis. Also, the specific economic challenges of the 1970s provide a good opportunity to assess the extent of regulatory change, the resiliency of the new regulation and the behavior of business.

In 1968, Senator Henry Jackson held a series of hearings on the environmental impact of surface mining. The Senate Committee on Interior and Insular Affairs took testimony from surface mining firms in various extractive industries, trade associations, environmentalists and government officials from all levels. These hearings focused on the need for a federal program, though no legislative proposals emerged.

From the Jackson hearings, which established the environmental hazards of surface mining, until 1971 little was done to correct many of the bad mining practices, especially in the Appalachian coal mining industry. Consequently, federal legislation appeared. A radical bill aimed at abolishing surface mining was introduced by Representative Ken Hechler (D-W.Va.). This bill, H.R.4556, made it out of committee, though it failed to pass in the full House. Such support for so radical a measure no doubt surprised the mining industries. Even the Nixon Administration introduced its own version of SMCRA, significantly watered down (S.993). These measures indicated that the 92nd Congress would take up the issue of surface mining regulation, if in less radical form than the Hechler bill.

In 1972, three more bills were introduced, H.R.60 by Wayne Hays (D.-Ohio), S.631 by Henry Jackson and H.R.6482 by the House Interior Committee. H.R.6482 passed the full House, but the 92nd Congress ended before the Senate could act.

In the 93rd and 94th Congresses a number of individual bills appeared, the most important of which were S.425 and its companion, H.R.25. In 1974, S.425 was passed and President Ford pocket vetoed it. Subsequently, in 1975, he vetoed H.R.25 as well with the Congress sustaining his veto by a three-vote margin.

By the seating of the 95th Congress and the accession of Jimmy Carter, the enactment of SMCRA was a virtual certainty. Early on each house reported a surface mining bill, H.R.2 and S.7. H.R.2 eventually became SMCRA and President Carter signed it into law in early August 1977. This entire legislative history and the subsequent years in which regulations were written and implemented mirror the salient features of the new social regulation and illustrate important shifts in government-business relations brought about by that regulation.

Structure of the Analysis

This research, through a case study of SMCRA, its enactment and its implications, attempts to develop an understanding of politics under the new social regulation. More specifically, it illustrates how the new social regulation can dramatically change the role of business firms in regulatory politics. In order to see how such changes occur, this study develops a model of firm behavior in regulatory politics from a synthesis of political and economic theory. The persuasiveness of this model is predicated on arguments that the individual firm is the primary business participant in regulatory politics. Second, the model is systematically applied to the case of SMCRA's evolution and implementation. The application of the model demonstrates how firms respond both externally and internally to SMCRA, how firms interact with other participants in regulatory politics such as trade associations, public lobbyists and legislators and which of these participants is likely to win or lose in the context of the new social regulation. Such conclusions, of course, constitute only working hypotheses for further investigation. Nonetheless, to the extent that the theoretical arguments are coherent and persuasive and to the extent that the model is faithfully applied to the case study, these will be strong hypotheses that can enhance our understanding of regulatory politics and business behavior therein.

2　Regulatory Politics: A Framework of Analysis

As a field of study, regulatory politics is far from uncharted. In constructing a framework of analysis we can in fact turn to a substantial body of literature, or, more precisely, two bodies of literature. Both political scientists and economists have devoted a great deal of effort to analyzing regulatory politics. Though each discipline displays characteristic strengths and weaknesses, taken together they can provide a relatively strong theoretical underpinning for the study of regulatory politics. This chapter assesses the strengths and weaknesses of political and economic theory in an effort to synthesize the two and specify a model of regulatory politics well suited to the new social regulation.

Political Theory

Wittingly or unwittingly political studies of regulation have stressed unity of interests and action on the part of business. Until recently only two widely cited works in political science emphasized cleavages among individual firms in regulatory politics. Those two studies, one by E. E. Schattschneider in 1935 and the other by Bauer, Pool and Dexter in 1963, examined how regulatory policy could lead to different preferences and behaviors among firms even within the same industry. Both works found significant variation across firms in terms of preference and behavior. Even more significantly, they found that they could account for this variation systematically—certain kinds of firms revealed particular preferences and behaviors.

Why did Schattschneider and Bauer et al. uncover interfirm differences in contrast to almost every other analysis of regulatory politics?

More to the point, why did they not begin with the assumption that business aggregates rather than individual firms were the relevant units of analysis? The answer lies in their choice of issue for study. Both research projects examined the politics of tariff policy.

In the most elementary terms one would expect firm preferences on tariffs to diverge between exporters and importers. On a more sophisticated level, we would want to know something about firm size and the proportion of operations affected by the tariff or the proportion of its operation in exporting and importing. While the effects of tariff regulation will naturally divide businesses into groups (possibly two groups, exporters and importers), there is not necessarily a predetermined way to order them. Each individual firm falls into one group or another based on the tariff effect. Under industry-organized regulation, on the other hand, it is likely that the industry group will have a common or at least a majority view on any particular policy. The nature of tariff policy, then, underscored interfirm differences.

In terms of its impact, the new social regulation bears a striking resemblance to tariff regulation. Neither type is industry-specific. Nor does either type fall under the jurisdiction of an industry-specific agency. Both are functional in their application and design; they apply more or less uniformly across industries and firms, imposing costs and benefits in a systematic way. In both cases these costs and benefits are imposed without regard to and, often in spite of, competitive considerations. These similarities suggest that a firm-level approach to the study of the new social regulation might be fruitful. The question, therefore, arises: why have views of business as a collective actor in regulatory politics so dominated political theory? Certainly the choice of a collective actor as a research focus requires substantial theoretical justification. Unfortunately, a number of historical, ideological and methodological factors have conspired to deflect political theorists' attentions from this important question.

The first factor that has led political theorists to view business participants in regulatory politics as collective actors is what may be termed our Madisonian heritage. Among the founding fathers, James Madison was the most articulate spokesman for what has become a venerable tradition in political thought, namely the presumption that group interests constitute the fundamental elements in political interaction. The constitutional system assumes that collective actors or "factions," to use Madison's term, vie with one another in the public pursuit of private interests. Madison, moreover, had no doubt that the fundamental source of political cleavage was economic interest. As Federalist #10 points out, "the most durable source of factions has been the verious [*sic*] and unequal distribution of property. . . . A landed interest, a manufacturing interest, a mercantile interest, a monied interest with many lesser interests, grow up of necessity in civilized nations and divide them into different classes . . ." (Madison, 1789, p. 79). Ostensibly society generally and business in particular is divided into many different collective actors. Thus, regulatory politics especially should reflect a contest among collective actors. The significance of this argument for our purposes is its powerful and persistent influence on contemporary political theory.

A second reason for the focus on collective actors in American business is the pervasive impact of the debate between pluralists and elitists. These two schools of political theory both see business as a dominant political actor in regulatory affairs. Their work focuses on a variety of participants from the private sector: an industry; the military-industrial complex; or simply business-writ-large. The important point is that each automatically aggregates business to a level above the individual firm. The pluralist-elitist debate helped to set the research agenda for political theorists including those interested in regulation. Furthermore, the vigor of the debate, stemming from strong underlying ideological perspectives, insured the persistence of the debate and its assumptions about business aggregates.

In contemporary political science the theme of group interests provides the basis for pluralist interpretations of the American political-economic system. If factions or groups form around interests, and if the political system is designed explicitly to prevent domination by any one group, much less business as a whole, it seems logical to construe competing groups as the crucial political actors. According to David Truman, a leading exponent of pluralist theory, "Many interest groups in the United States are politicized. . . . Both forms and functions of government, in turn, are a reflection of the activities and claims of such groups" (Truman, 1951, p. 11). If business interests participate in regulatory politics, pluralist theory views them as it does any other legitimate interest. In *The Group Basis of Politics* Earl Latham sets forth the pluralist view of politics in America: "The legislature referees the group struggle, ratifies the victories of the successful coalitions, and records the terms of the surrenders, the compromises and conquests in the form of statutes. . . . What may be called public policy is the equilibrium reached in this struggle at any given moment . . ." (1965, pp. 35–36).

In the pluralist perspective business groups must vie with other groups in the political system. While business may seek governmental aid, more often than not business tries to ward off attempts of other collective interests to regulate economic activity. More important, perhaps, business as a whole is unlikely to have a common interest around which to organize on particular issues. Not surprisingly, pluralists take a rather sanguine approach to business influence. The central government, in their view, is a significant environmental factor with which business must cope in generating national wealth. Ultimately pluralists have little reason to fear political domination by business since our political institutions were specifically engineered to cure the mischief of factions.

In contrast, elitists, whether of a loose populist strain or the more orthodox Marxist variety, see business interests as inimical to the general welfare. Rather than viewing various business interests as relevant political actors, elitists are inclined to assume that the private or capitalist sector of the economy comprises an element in liberal democratic society. Therefore business, at least big business, constitutes a unified collective actor. A more refined version of this argument postulates an important difference between monopolistic or oligopolistic business and small business, the former being the object of concern (see, for example, Kolko, 1965, Baran and Sweezy, 1966, or Mills, 1959). Nonetheless, the thrust of both is to take business as a political-economic factor in and of itself. From this assumption elitists argue that business' unique position in the political economy enables it to dominate society: "The market power which large absolute and relative size gives the giant corporation is the basis not only of economic power but also of considerable political and social power of a broader sort" (Kaysen, 1975, p. 93).

The obvious contradiction between this view of the political economy and the purportedly democratic basis of society as well as pluralist theory accounts, in large part, for the tenacity with which elitists have maintained other critiques of business in America. As Gabriel Kolko put it: "The real questions are (1) Do a small *group* of very wealthy men have the power to guide industry and thereby much of the total economy, towards ends that they decide are compatible with their own interests? (2) Do they own and control major corporations? The answers must *inevitably* be affirmative" (1965, p. 69). This portrait of American business accounts for democratic demands on the central government to promote the general welfare through regulatory policy. Irving Kristol, certainly no friend of the elitist perspective, suggests in this regard that

"A Democracy is not likely to permit huge and powerful institutions (i.e., big business) with multiple spillover effects on large sections of the population to define its interests in a limited way" (Kristol, 1976, p. 40).

The gravity of the elitist allegations, the time-honored lineage of the pluralist argument as well as the poignancy of their debate all served to focus political analysis of regulatory policy on collective actors, especially in American business. With hardly a thought about theoretical validity, political theorists assumed that business aggregates rather than individual firms were the appropriate units of analysis in regulatory politics.

Unfortunately, even as the energy and enthusiasm of the pluralist-elitist debate was waning, another school of political thought, which also focuses more or less uncritically on business aggregates, arose and gained currency among political theorists. This new perspective, known as "plural-elitism" or "interest group liberalism," put forth a persuasive description of regulatory politics. Interest group liberalism had a strong appeal since it seemed to effect a reconciliation of the pluralist and elitist positions. Interest group liberalism postulated that the emergence of positivist government and the administrative state in the twentieth century fostered an environment in which there is political competition, but only among permanent, organized interests. This, of course, departed from the earlier pluralist view that saw interest groups as shifting and ephemeral. It also differed from elitism in acknowledging that any organized interest could be an effective participant—naturally, though, this regulatory environment may well favor big business with its legal, financial and organizational resources.

According to interest group liberalism regulation is neither the product of pluralist interaction among coequal competitors nor the reflection of a political economy dominated by big business. Rather, regulation, like any other public policy, is a function of organized interests that predominate on the contemporary scene. As Theodore Lowi explains: "Congressmen are guided in their votes, Presidents in their programs, and administrators in their discretion by whatever organized interests they have taken for themselves as the most legitimate; and that is the measure of legitimacy of demands" (1969, p. 72). In such a system ossification, deadlock, inflexibility and a sense of mere democracy rather than true representation, are the byproducts. This theoretical perspective in political science dates back at least to the classic study of policy making in postwar America, *Muddy Waters: The Army Engineers and the*

Nation's Rivers (Maas, 1951). In this analysis, Arthur Maas demonstrates how organized interests predominated in a policy-making subsystem: rather than Congress as a whole, particular committees or subcommittees provide the political arena.

This portrait of policy making gave rise to the well-known concept of "iron triangles." Since the 1960s, however, the concepts of "subgovernments" (see Ripley and Franklin, 1980) and "issue networks" (see Heclo, 1978) have become the stock in trade of political scientists studying regulatory policy. These newer concepts suggest a more open policy process, though one still fixed, fragmented and dominated by organized interests.

While interest group liberalism represents a significant advance over pluralist and elitist theory, if for no other reason than it moved beyond the ideological debate, it too operates on the assumption of group or collective actors. Business participated in regulatory politics effectively only in organized interests. That business found it easier to organize could explain the success if seemed to have in the contemporary political milieu. But it was organized business interests that were the relevant political actors.

Beyond the philosophical and ideological concerns about the role of business in a liberal democracy, political science has concentrated on business aggregates for pragmatic reasons in the construction of empirical theory. As a practical matter, regulation of business by the federal government has been organized generally on an industry-by-industry basis. Government regulates the oil companies, the railroads, the communications industry, etc. Political theorists tended to view entire industries as unified and single-minded actors.

Leading works on regulatory politics clearly reflected the notion of an industry as a political-economic actor: "With the exception of industries like radio broadcasting, air transport and trucking which demand regulation to restore order in a chaotic situation, *groups subject to regulation* have always fought against the adoption of public regulatory policies" (Bernstein, 1955, p. 251, emphasis added). Or "Thus each agency develops a *constituency of its own, which is the industry it was created to regulate*" (Kohlmeir, 1969, p. 9, emphasis added).

Another practical factor fostering the group approach to government regulation is the existence of trade associations to represent business in arenas of regulatory politics. While these trade associations were not a product of industry-based regulation, they reinforced the tendency of researchers to focus attention on industries as political-economic actors.

Pluralists could find support for their view in the behavior of specific trade associations. More general associations such as the National Association of Manufacturers (NAM) or other broad-based organizations like the U.S. Chamber of Commerce provided elitists with examples of higher order business aggregates. Highly visible in regulatory politics, these associations and organizations attracted the interest of political scientists. Typically, one discovers such assertions as, "By means of their economic power, their familiarity with the channels of government decision-making, their knowledge of the details of policy issues . . . their ability to organize and orchestrate the views of their members, vested interests are able to do what they will" (Haveman, 1973, p. 6). This kind of statement indicates that associations which develop around economic interests can effectively organize political action. Again this suggests a view of business as a collective actor or set of actors.

In their public behavior these actors appear unified and harmonious in their views on regulatory policy or government-business relations. The obvious methodological and theoretical appeal is that researchers need not concern themselves with discovering how to aggregate business behavior. From the behavior of collective actors, the common interests of the industry or business groups may be inferred.

The Problem with Political Theory

Although pluralists, elitists and interest group liberals have had tremendous influence on the study of government-business relations and regulation, important criticisms can be leveled against their work, especially as it relates to newer social regulation. Empirical studies indicate that a firm-level approach may be more apt than an approach at a higher level of aggregation. In deductive, theoretical, terms as well there are powerful arguments for concentrating on the individual firm.

Taking the empirical work first, a number of recent studies suggest that the firm-level findings of Schattschneider and Bauer et al. have applications outside the issue area of tariff policy. For example, some students of regulatory politics, rather than taking trade associations or broader business organizations to represent unified collective actors, have probed beneath the surface of formal organizations. Their investigations point to the significance of firm preferences and behaviors. For example, a study of the home appliance industry and its trade association found "that many major firms differed with respect to their economic position and internal organization" (Hunt, 1978, p. 126). Analogously to the two earlier political science studies on tariff policy, this research concluded that these interfirm differences related systematically to preference and behavior.

In a similar vein, Edwin Epstein argued that "corporations often have political interests sufficiently limited and funds sufficiently ample that they need not be unduly committed to associational policies" (Epstein, 1969, p. 102). It appears that formal business organizations may mask a good deal of interfirm diversity. The results of Bauer et al. offered this insight over twenty years ago: "Cautious procedures are employed for reaching a policy position, and spokesmen are confined to stating that position without elaboration, for fear that even the most cautious elaboration may produce dissension" (1962, p. 337). That these considerations seem so plausible and yet had such little impact on political science testifies to the pervasive influence of the elitist, pluralist and interest group liberal perspectives.

These three dominant perspectives received another challenge from Gideon Doron's study of regulating the tobacco industry. The significance of Doron's challenge to the traditional assumptions concerning aggregate business interests may not be readily apparent since his study demonstrates the effectiveness of the tobacco industry in thwarting social regulatory efforts. However, Doron does not assume collective action, much less collective interest.

In a key theoretical passage of *The Smoking Paradox* he asserts: "It is assumed that a 'firm' is a profit maximizer and that it will be risk averse when faced with a threat. When an external source threatens the profit margin or even the existence of the firm *and* the industry, one would expect the industry to form a 'grand coalition' or 'a coalition of the whole'" (Doron, 1979, p. 31, emphasis added). Although the cigarette industry formed a "grand coalition," individual firms made independent judgments about collective action. In this case it so happened that the profits of all firms in a particular industry were threatened by an external source. But what if we change the scenario such that a regulatory standard is applied uniformly to individual firms in different industries and/or in vastly dissimilar economic circumstances, a situation that typifies the new social regulation? The variation in circumstances (i.e., production functions) may threaten some firms but hold out promise of economic rewards to others. Recall the Smead observation that regulation hurts some firms and helps others. Doron's framework of analysis logically implies that a "grand coalition" would not form—it is conceivable that smaller coalitions of similarly situated firms would coalesce into unified political participants.

Several smaller coalitions are exactly what Bruce Ackerman and William Hassler found when they investigated the impact of clean air regulation. Indeed, this new regulation gave rise to undreamed of political coalitions as it systematically imposed costs and benefits on coal firms and public utilities as well as industrial users of various energy sources. The new social regulation, as Ackerman and Hassler demonstrate, occasioned serious intra-industry political cleavages (1981).

If inductive and empirical research has cast some doubt on political theory, certain deductively based arguments have provided an extremely powerful case for a firm-level approach to regulation. Empirical work has shown that the assumption of common interests among firms in a group, industry or wider, is open to question. In his analysis of collective action, Mancur Olson delivered a more telling blow by showing that common interest is not sufficient to insure that an aggregate of firms will behave as a unified actor. Attacking Marxists as well as group theorists, he demonstrated that collective action can be understood only in terms of the motivation of individual group members (Olson, 1965).

A common ground for elitism, pluralism, interest group liberalism or firm-level theory is the assumption that political-economic actors, no matter the degree of aggregation, are self-interested. Accepting this premise, Olson questions whether there is any logical reason to believe that an aggregation of firms will act together to achieve a collective purpose. His answer is that even if a broad common interest exists, individual firms are unlikely to act in concert. Olson asserts that elitist theorists are mistaken: "Class action (bourgeois or proletarian) will not occur if the individuals that make up the group are rational" (p. 105). Further, he rejects pluralist arguments because "They generally take for granted that the individuals in these groups will act to defend or advance their group interest" (p. 126). His reasoning is straightforward. If self-interest motivates individual action, groups must form for the benefit of individual members. This point applies with equal force to interest group liberalism.

However, even the fact that a number of actors, for example, firms, share an interest will not guarantee group formation or collective action. Because joining a group and supporting its activities imposes costs on members, each self-interested firm weighs the costs of joining against the benefits of collective action. Moreover, if the benefits are public goods, equally distributed and impossible to deny to any one member, the number of individuals sharing an interest in a public good becomes crucial. In other words, the potential size of the aggregation is an important determinant of the likelihood of collective action.

Relatively small groups of firms are more likely to cohere and act because the benefit of collective action may be sufficient to induce at least a few members to bear the costs. For intermediate or larger groups, argues Olson, coercion and distribution of selective benefits through the organization provide the only explanations for collective action. In a sense, he develops a theory of collective inaction.

Olson's most significant point, in terms of this research, is that it makes more sense to begin with an internal analysis of voluntary organizations such as trade associations or socioeconomic groupings such as "big business" than to consider the behavior of those aggregate actors (i.e., to assume collective action at the outset). To look on an aggregation of firms as a single, self-interested actor imputes a certain homogeneity among them on a particular regulatory issue or set of issues. While this may be the case, there is no reason to accept it a priori. And, more important, individual costs of collective action will inhibit groups. If aggregate actors do emerge, there will be a firm-level basis to them. This, of course, is precisely Doron's theoretical point in *The Smoking Paradox*.

Economic Theory

Unlike their counterparts in political science, economists studying federal regulation have developed firm-level theory. Such theory is most compatible with the prevailing paradigm in economics: the rational, self-interested firm constitutes a primary behavioral unit in microeconomic theory. For a variety of reasons, some epistemological and some professional, political theorists ignored, at least into the late 1970s, economic studies of regulation. Admittedly the bulk of economic theory dealt almost exclusively with the behavior of public utilities under rate regulation. This work focused on differences between firm behavior under free market conditions versus regulation (see, for example, Averch and Johnson, 1961, or Becker, 1962). While these issues are not central to political theorists, economic theory does suggest the possibility that the study of regulation might be based on firms rather than groups of firms.

Of direct concern to political theorists, though, is the work on regulation initiated by George Stigler at the University of Chicago. In using economic assumptions and models to analyze a political phenomenon, namely regulatory policy, Stigler builds on earlier economic theories of political behavior. In particular, he owes a debt to Anthony Downs and Mancur Olson whose *Economic Theory of Democracy* and *Logic of Collective Action* respectively have brought models of rational decision making and economic behavior to political theory.

In his work, Stigler, like Olson, allows for group behavior, though beginning with the firm as the behavioral unit (Stigler, 1972). He observes that government can bestow benefits as well as impose costs on firms through its regulatory powers. Moreover, he argues that because regulation is a good, albeit a public good, there will be a demand and supply for its provision. Specifically, firms should demand government intervention in order to acquire various economic benefits which government may bestow through its regulatory powers. According to Stigler, then, the demand for federal regulation originates with rational, self-interested firms.

As Richard Posner notes in his excellent review of the regulation literature, this conclusion is similar to that of some political theorists (1974). In particular, Gabriel Kolko and Marver Bernstein as well as some in the interest group liberalism camp make similar arguments about the origin of regulation. Kolko suggests, for example, that the Interstate Commerce Commission (ICC) and the Food and Drug Administration (FDA) were created to satisfy the demands of railroads and major meatpackers respectively (1963). On the other hand, business also may co-opt already established regulatory institutions for its own benefit. Bernstein develops this line of argument, suggesting a life-cycle theory of regulation under which industries "capture" a regulatory agency over time as public interests and involvement in an issue wane (1955). In a similar vein, interest group liberalism postulates that organized interests operating in discrete subgovernmental networks demand governmental largess (see, for example, Reich, 1964, or Fiorina, 1977).

Despite his ostensible similarity to these political theorists, Stigler differs in an important respect. He begins with the firm as the unit of analysis and bases his conclusions on an identifiable body of theory, namely microeconomics, rather than intuitively appealing but non-rigorous assumptions. Consequently, Stigler's work is conceptually and theoretically more precise. Although he arrives at conclusions similar to those of some political theorists on the subject of regulatory origin, his deductions are based on theory rather than common sense.

Specifically, Posner explains, the microeconomic theory of cartels,

Olson's approach to problems of collective action, underlies Stigler's theoretical framework:

> Since the effect of typical regulatory devices is the same as that of cartelization—to raise prices above competitive levels—the benefit side of cartel theory is clearly relevant. The cost side also seems relevant. The members of an industry must agree on the form of regulation. And just as the individual seller's profits are maximized if he remains outside the cartel (as long as his competitors remain inside) so any individual firm that would be benefitted by a type of regulation will have some incentive to avoid joining in the efforts of its group to obtain the regulation. (1974, p. 644)

As Olson pointed out, these costs can be mitigated by small numbers, homogeneous interests, coercion and/or selective benefits. Thus we may find concerted regulatory participation from business, but it has a basis in the behavior of individual firms. Firm-level theory also can account for non-cooperative political action from firms. In other words, it is a superior theory because it accounts for a greater percentage of behavioral variation.

The Problem with Economic Theory

Although economic theory offers a more rigorous and satisfying explanation for business behavior in regulatory affairs than political theory, it too suffers from a serious weakness. While it accounts for demands for regulation emanating from business firms, it is far less successful in explaining what happens when these demands filter through the political system. On the supply side, Stigler's theory degenerates into assumption and speculation no more rigorous than political theory on the demand side. Understanding the supply side of regulation entails, at a minimum, a description of relevant political actors and a theory of political influence. Without this supply theory it is impossible, in Posner's words, "to explain the pattern of government intervention in the market" (Posner, 1974, p. 656).

Sam Peltzman, an associate of Stigler's, has attempted to develop a framework with which to analyze the political-economic process that generates federal regulation. Peltzman characterizes regulatory institutions as a tool with which congressmen and bureaucrats can conspire to direct government benefits to their respective constituencies which demand regulation. These constituents offer votes or campaign contributions to legislators and other forms of political support to bureaucrats in exchange for political favors on regulatory matters (Peltzman, 1976). Thus, Peltzman depicts the supply of regulatory policy by imputing rational self-interest to political institutions as well as business institutions.

This approach resonates with interest group liberalism in suggesting a political system that responds to organized interests, though in Peltzman's version firms may act individually as well as collectively. While Peltzman's framework represents an advance over Stigler's theory, it is not satisfactory either. It vests far too much positive control over regulatory policy in the Congress and regulatory agencies. In addition, it oversimplifies the workings of Congress and its interactions with the bureaucracy.

Research in public policy indicates that federal regulation is "supplied" through a complex of interaction of political-economic forces. Noll and Fiorina, 1976; Ripley and Franklin, 1980; Wilson, 1980). Regulatory politics as envisioned by Peltzman, though, entails a narrow field of action and a limited cast of characters resembling the so-called "iron-triangle" model. In this view, much as Peltzman asserts, regulatory policy is the product of a closed tripartite interaction among agency, constituency and key legislative committees.

This model of regulatory politics may be compatible with later stages of some kinds of regulation in which legislators and bureaucrats hold considerable leverage. However, it does not square with scenarios for early stages of development in regulatory programs or attempts to reorient or reform regulatory policy. In these wider settings much more is involved than legislators and bureaucrats serving as conduits for businesses "demanding" regulation. In these wider settings those demanding regulation are unlikely to be business firms, as economic theorists as well as elitists and some interest group liberalism theorists would assert. Indeed, looking at the broad sweep of regulatory policy and its development, it seems that popular demand for government intervention has played a significant role. It should be evident to the most casual observer, for example, that the spate of environmental and consumer regulation in the 1960s and 1970s was not initiated by business demands.

Also the peculiar nature of the new social regulation—functional rather than industry-specific organization—clearly puts it outside of Peltzman's narrow theory. Those demanding the new regulation as well as the complex political economic forces involved resemble more the pluralist model of government responding to cross-cutting political demands. As Hugh Heclo argues in assessing regulatory theories such as Peltzman's: "Control is said to be vested in an informal but enduring series of 'iron triangles' linking executive bureaus, congressional committees, and interest group clienteles with a stake in particular programs. Looking for closed triangles of control we tend to miss the fairly open networks of people that increasingly impinge upon government" (1978, p. 57).

This political complexity must be brought into any serious analysis of federal regulation. One way to accomplish this is to cast firm behavior in subgovernmental settings. Subgovernments, according to Ripley and Franklin, may be characterized as: "small groups of political actors, both governmental and non-governmental, that specialize in specific issue areas" (1980, p. 12). The influence that a firm or group of firms can wield over regulatory policy is largely a function of the particular subgovernment in question.

Though similar to Peltzman's view of regulatory politics, the subgovernmental model differs in two important respects. First, it allows for a much wider cast of organized interests and governmental actors. Second, subgovernments are open to significant outside political pressures including broad-based demands for regulation. This is crucial in understanding broad regulatory change and, as we shall see, the new regulation. The first point draws on the interest group liberalism view of regulatory politics, whereas the second point draws on pluralist theory. Both can contribute to our understanding of regulatory supply and demand.

Integrating Political and Economic Theory: A Model of Regulatory Politics

The foregoing assessment of political and economic theory suggests that any model of regulatory politics ought to reflect at least three ideas. First, individual business firms rather than collective business actors ought to figure prominently in the framework. Second, these firms (as well as any other participants in regulatory politics) must be construed as rational, self-interested actors. This assumption, of course, implies a model of firm behavior which must be set forth. Third, the framework ought to take account of the considerable political-economic complexity within which federal regulatory policy is enacted and implemented. What follows is an attempt to integrate each of these three conceptual elements.

Just as economists theorize about a firm's rational behavior in much the same way as they do about a consumer's, political scientists can consider a firm's rational behavior analogously to a citizen's. Individual business firms as well as people participate rationally in both economic and political life. Rational participation in either case entails the conscious and calculated connection of individual preferences to specific behavioral alternatives. How, though, is this connection effected?

Designating a firm as a rational actor implies only a general behavioral model. In the loosest sense it means a firm acts for reasons. Under a more restrictive definition, it means a firm acts for reasons only after some careful consideration. Their reasons, in other words, must be based on calculation rather than passion. This precludes psychoanalytic models of behavior—even Freudian explanations can find "reasons" for action.

More specifically, rational behavior is purposive. Business firms relate means to ends, actions to goals or, in the more formal language of rational choice theory, behavior to preferences (see Luce and Raiffa, 1957, and Riker and Ordeshook, 1973). A rational firm will select, from a set of alternative participatory behaviors, that behavior which will bring about its most preferred outcome from the set of alternative regulatory outcomes.

A rational firm, moreover, will select the least costly behavior for achieving its preference. Cost, therefore, will be balanced against effectiveness. This being the case, the resources at a firm's disposal delimit both its behavioral alternatives and its effectiveness. Plentiful resources in terms of money or political and technical expertise set the parameters for behavioral alternatives open to a firm. Collective action is but one choice among these alternatives. The critical determinant of choice, though, remains preference—costs of participation in regulatory politics simply qualify the impact of preference. Thus, the first step in understanding firm behavior must be to develop a concept of firm preferences.

Of course it is possible to break the firm down into its organizational elements, arguing that the incentives and preferences of each pull the larger entity in different directions. For our purposes, though, such reductionism would not be useful. Our attention would be deflected from the larger issues of regulatory politics to the study of the inner workings of the firm. There is, moreover, ample justification from the literature in microeconomic theory for treating the firm as a unitary, rational actor.

Microeconomic theory postulates that the individual firm seeks to maximize a particular goal (π) expressed as a function of price (p), quantity of a good (q), variable cost (VC) and fixed cost (FC). Economists model firm behavior by the expression: $\pi = pq\text{-}(VC + FC)$ or the difference between total revenue and total cost. This equation represents the firm's profit function. The profit function provides a means of distinguishing among firms since it will vary from firm to firm. The reason for this variation is that different production functions underlie particular profit functions. Specifically, a production function depicts the relationship between amounts of different kinds of inputs and total output, that is, the relationship between fixed and variable costs on the one hand and quantity on the other. Clearly the profit function embodies this relationship. Just as clearly it will differ among firms as the level and age of technology differs among them.

Variations in production and profit functions can lead to different preferences among firms with respect to regulatory policy since each firm, acting rationally, interprets exogenous impacts in terms of their effect on its own profit function. Federal regulation is such an exogenous impact. To the extent that federal regulation impinges on profit maximization, the firm will form preferences with respect to that regulation. Those preferences, in turn, will guide the firm's behavior.

At this juncture, we can depict a rudimentary model of firm behavior in regulatory politics. The profit function serves as a filter through which the firm interprets external stimuli, in this case regulation. Through its interpretation, the firm develops preferences which guide its behavior. This model suggests that firms will oppose regulation that decreases net revenue and, as George Stigler and other economic theorists argue, support regulation that will increase net revenues.

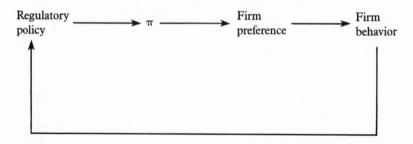

Figure 2.1

In applying this model of firm preferences and behavior, one might predict, for example, that public utilities would "demand" rate regulation or airlines would "demand" federal assignment of routes because experience has shown a positive net revenue effect. Conversely, one might argue that rational manufacturing firms would oppose environmental regulation since it significantly increases fixed costs. However, this intuitively appealing argument is flawed. The impact of environmental regulation, increasing fixed costs, may be offset in the longer run by other advantages. Simple predictions or explanations on the basis of a profit function cannot explain all firm behaviors. Some firms have allied themselves with consumer advocates or environmentalists on issues of regulatory policy. How can one account for this without abandoning π as a determinant of preferences?

In an important discussion, Cyert and March (1963) have suggested that firms seek to maximize price per share of stock rather than π. Yet profit constitutes a constraint in that some acceptable level of profit, if not profit increases, determines stock price. An alternative goal for the firm to maximize, argues Baumol (1967), is aggregate sales. Under this interpretation, the firm is most fundamentally interested in total sales. Profits provide a constraint in this case as well since, in the medium to long run, revenues must exceed costs for the firm to survive.

Robin Marris attempts to show that growth is the objective of firms under a "theory of managerial capitalism." However, he acknowledges: "if capacity in terms of gross assets were to expand more rapidly than the volume of saleable output, average capacity utilization must decline and with it rate of return (i.e., π). But a declining rate of return cannot . . . be reconciled with a constant growth rate" (1964, p. 118). In each case of economists trying to bring firm behavior in line with reality, profit remains an underlying constraint.

The profit function then is at the core of more complex considerations. Different firms may have different time horizons or may plan on different criteria. Firms also may vary in their access to financial markets. An environmental regulation will increase fixed costs for a firm but also may raise the possibility of increased aggregate sales, higher stock price or larger market share in the long run. A firm with relatively high fixed costs may view regulation differently than one with relatively high variable costs and low overhead. Therefore, when this research uses the profit function to model firm preference formation it should be understood that more complex considerations are involved than simply immediate net revenues.

There is a second important qualification to this model of firm preferences and behavior. As in any model of rational choice, problems of risk and uncertainty arise. Based on internal criteria, a rational firm will prefer some particular outcome of the regulatory process to others on any issue. Indeed, we may assume that a firm has a preference ordering for all conceivable outcomes. If a firm prefers one outcome and "knows" that a particular behavior will lead to that preference, it will choose that behavior. Explaining or predicting firm behavior under these circumstances becomes a theoretically trivial exercise.

A problem, of course, arises in that the world of political economy is not certain. Neither we nor a firm can "know" that a behavior will insure a preference. Political economy is a risky environment in which a firm must estimate for itself the chances of any behavior leading to its preferred outcome. At least in a rough and ready way, the firm must make three calculations. First, the firm must decide the chances of a behavioral alternative bringing about its preferred outcome as opposed to any other. Second, it must decide on the probability of its preference resulting from that behavior rather than another. Finally, it must decide whether to act at all based on the above calculations.

Even if one behavior is the most likely of any to result in the preferred outcome, should its probability of succeeding be relatively low, the firm may choose not to act. The choice of acting or not also depends on the value of the preferred outcome to the firm. Thus, if the probability that a behavioral alternative will yield the preferred outcome is low but that outcome has tremendous significance, the firm may act nevertheless. The firm determines its behavior through an expected value calculation, that is by considering both value and probability of success (Riker and Ordeshook, 1973).

In developing an explanation of firm behavior in regulatory politics we can conceive of the firm as a rational actor behaving on the basis of (a) preference for a particular social outcome, (b) the value of the preferred outcome to the firm, (c) its determination of the chances that a particular behavior will achieve that preferred outcome.

Preference formation, though, is only the first step through which a rational firm generates a response to regulation or proposed regulatory change. The firm must associate its preferences with behavioral alternatives, taking account of costs for these alternatives. It then must choose a course of action. This "calculus of participation" includes an assessment of risks as well as costs.

Firm behavior and our analysis of it both depend on an understanding of the relevant subgovernment in which regulatory policy will be enacted and implemented. Accurate perception of the subgovernmental setting is essential for the firm's rational calculus of participation, not to mention its success in pursuing its preferences. As James Post maintains in his study of corporate behavior: "Organizations of all sizes and types face the challenge of operating in a social setting that is increasingly complex and inherently political. The manner in which organizations respond to this commercial and social complexity is fundamental to their institutional legitimacy and their business survival" (1978, p. 5). The complexity and politics of Post's "social setting" is brought directly to bear on business firms in the subgovernmental context.

The subgovernment setting of regulatory policy places the simple microeconomic model of firm behavior in a more complex political-economic context. Thus, the individual firm assesses regulatory policy in terms of its impact on the profit function, forms preferences based on that assessment and selects its behavioral response subject to the subgovernmental constraints on its probability of influencing regulatory policy. Figure 2.2 depicts this more detailed model of firm behavior.

Preferences formed on the basis of π are causally related to social outcomes via firm behavior (Nagel, 1975). This connection is subject to an intervening variable, namely the firm's estimate of its probability of success as determined by its perception of the subgovernmental setting. This setting combines with firm behavior to determine social outcome. Social outcomes can have an effect on the subgovernmental setting to the extent that regulation may change patterns of political influence.

This model of regulatory politics suggests that insofar as social out-

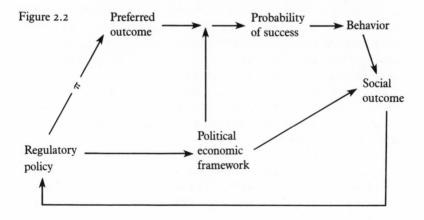

Figure 2.2

comes conform to firm preferences, that firm has exercised influence. However, the empirical analysis of influence relationships requires more than noting a correlation between preference and outcome. It demands a careful qualitative analysis of influence processes. It also necessitates paying close attention to the mechanisms and structures through which a firm may exercise influence on regulatory policy.

Conclusion

This critical evaluation of political and economic theories on regulatory policy indicates that each discipline exhibits particular defects. Political theory, with few exceptions, automatically imputes a community of interest and collective action to industries or still larger aggregations of business firms. Both empirical and purely theoretical considerations cast doubt on this formulation as a basis for understanding regulatory politics in general. The following two chapters will demonstrate that the assumption of collective business behavior is particularly ill-advised in the context of the new social regulation.

Although economic theory is more cogent in this regard—it sees individual firms as the primary behavioral units even if collective action materializes—it too proves unsatisfactory in its attempt to comprehend regulatory politics. Specifically, it oversimplifies the political environment in which regulatory policy is made. Indeed it seems designed to explain business behavior in the context of preexisting regulatory programs with well-defined relationships among politicians, bureaucrats and businessmen. Such a depiction provides little insight into the early phases of regulatory politics when issues are defined and when a variety of political participants plays an active role. Nor does economic theory offer a compelling model of regulatory politics under the new social regulation. As political scientists have shown, subgovernments in this new regulatory environment are better characterized as permeable issue networks than as tightly knit iron triangles.

In view of the fact that the weaknesses of each appear to be offset by the strengths of the other, integrating political and economic theory should lead to a superior conception of regulatory politics, particularly under the new social regulation.

For any regulatory policy, we along with the rational firm can conceive of a range of behavioral responses. The choice of response reflects not only firm preferences, resources and costs of participation, but also the firm's view of the relevant subgovernment. Any issue of regulatory policy brings into play different actors, each in pursuit of its own self-interest. Depending on the regulatory policy in question and on the stage of the policy process, a firm may have to contend with legislators, judges, bureaucrats, other firms, trade associations, public lobbies and the media. Its influence, therefore, will be a function of the configuration of the regulatory subgovernment. Nevertheless, the individual firm and its calculus of participation remain the focus for our model of contemporary regulatory politics.

3 The Political-Economic Framework

What different sorts of behavior arise from the preferences of different firms in the coal industry? To answer this question requires us to take account of the political-economic context in which SMCRA was enacted and implemented. Beyond individual preferences, firm behavior depends on assessing the probability of exercising political influence in a given context. Consequently, our analysis must rely not only on an elaboration of firm preferences, but also a discussion of the new social regulation, including the relevant actors and institutions involved with SMCRA. While this complicates the analysis, it also enriches it, providing insight into the decisions of firms to participate individually, collectively or not at all. Delineating the political-economic framework confronting coal firms in the 1970s is the express task of this chapter.

Regulatory Change and Regulatory Regimes

During the last decade and a half we have witnessed a major shift in regulatory politics and policy. Analogous in import to the Progressive Era and the New Deal, this shift has been characterized as the "new social regulation." Implied by this term is a sense that beyond simply a third wave of regulation, the substance of regulatory policy has changed. In this regard, many observers have noted that regulatory policy has shifted from controlling marketplace behavior such as pricing and competition to insuring public health and safety (see, for example, McAvoy, 1979b). While some Progressive and New Deal policy dealt with these "social" issues, the new regulation has come to focus on them.

This chapter illustrates the character of the new social regulation by contrasting it with the two previous periods of regulatory growth. Further, it will demonstrate that the new regulation has fundamentally altered politics as well as policy. Among other impacts, the newer regulation has accentuated the individual firm as a political participant while undermining the role of trade associations and other business aggregates. Specific examples will be drawn from SMCRA to show how it reflects the new regulation. Before we proceed with an analysis of the new social regulation, however, it will be useful to specify what is meant by a shift in regulatory politics and policy.

One way of conceptualizing such a shift is to think in terms of distinct "regulatory regimes." Generally, a regime is a system of administration or governance characterized by a specific configuration of institutions and actors and bound together by legal and political ties. A regulatory regime, therefore, may be understood as any such configuration dealing with regulatory policy. Naturally it will reflect the larger political regime. Yet, it may be thought of as a distinct set of actors, institutions and the relations among them. More precisely, a regulatory regime embodies: (a) a set of ideas justifying regulatory policy; (b) a set of actors in regulatory politics; (c) a set of institutions that develop and implement regulatory policy; (d) a set of regulatory tasks that government must perform. A new regulatory regime logically means new ideas, new actors, new institutions and new tasks. Major regulatory shifts, then, ought to be identifiable in terms of shifts in these four elements. The crucial element, though, the one that must be present for a regime change, is the emergence of new ideas. This sets qualitative regulatory change apart from mere growth.

New Ideas

The emergence of new regulatory regimes depends, in the first instance, on important shifts in ideas about government-business relations. The new social regulation brought such ideas to the fore in the 1970s. The conception of government-business relations underlying environmentalism especially reflected a sharp discontinuity with previous regulatory rationales.

In extending regulation, both Progressives and New Dealers hoped to correct market imperfections, restore competition and curtail monopolistic practices (Kohlmeir, 1969). In describing New Deal regulatory agencies, for example, William O. Douglas wrote: "They have become more and more the outposts of capitalism; they have been given increasingly larger patrol duties lest capitalism, by its own greed, avarice or myopia destroy itself" (cited in Bernstein, 1955, p. 25). Progressive and New Deal regulation was intended to restore the market, or at least its results.

Environmentalists, on the other hand, have demonstrated relative indifference if not outright antipathy toward the concept of a free market. Those who would characterize environmental regulation as simply a shift from price and entry control to pollution control miss a key point. Many environmentalists would consider the health of the market an ancillary concern at best. Some would argue, in fact, that a capitalistic economy has created and continues to exacerbate ecological problems. Perhaps it is best to turn directly to the words of environmentalists to convey their ideas.

In September of 1969 the first Earth Day was heralded by a *Declaration of Interdependence*. This document embodies the new ideas of environmentalism and sets it off in marked contrast from earlier regulatory periods. In a pastiche of the Declaration of Independence, this new document proclaimed:

> nature has instituted certain principles for the sustanence [*sic*] of all species, deriving these principles from the planet's life support system . . . that whenever the behavior by the members of any one species becomes disruptive of these principles, it is the function of the other members of that species to alter or abolish such behaviors and to reestablish the theme of interdependence with all life. . . . Prudence, indeed, will dictate that cultural values long established should not be altered for light and transient causes, that mankind is more disposed to suffer from its asserting a vain notion of independence than to right themselves by abolishing that culture to which they are now accustomed (cited in Caldwell, 1975, p. 3).

This excerpt leaves no doubt as to the iconoclastic ideas underlying environmentalism: capitalism figured prominently among the "cultural values long established" that had to be forsaken. Down with individualism and up with collectivism.

That these underlying ideas had much more than a rhetorical impact was affirmed by Bruce Ackerman and William Hassler in their perceptive study, *Clean Coal/Dirty Air*. They asserted: "The passions of Earth Day have marked our law in deep and abiding ways. Statutes passed in the early 1970s did more than commit billions of dollars to the cause of environmental protection in the decades ahead. They also represent part of a complex effort by which the present generation is revising the system of administrative law inherited from the New Deal" (1981, p. 5). This complex effort reflects a political economic outlook which relegates the concept of a free market to a romanticized vision of bygone days. In addition it reflects a frank acknowledgment of the failure of earlier regulatory efforts, at least in the eyes of those demanding the new social regulation. As the interest group liberalism character of the policy process became apparent in the post–World War II era, environmentalists and other advocates of the new social regulation changed their political strategies and tactics.

Accepting the interest group liberalism view of regulatory politics, they developed a new approach to regulatory policy. Andrew McFarland systematically depicts this approach, which he terms a "theory of civic balance": "This system of beliefs implies that widely shared interests are not adequately represented . . . unless citizens form new institutions for representation; American government will have an elitist character, in that economic, political and bureaucratic leaders will control public policy for their own benefit" (1976, p. 7). This concept of civic balance clearly captures the fear of corporate dominance, a felt need for transcending the market and the desire for participatory democracy, all characteristic of the new social regulation.

In both its legislative and implementation phases SMCRA bears the imprint of environmentalist ideas about the market and regulatory politics.

Environmentalist disregard for the potential anticompetitive effects of this outlook appears in the Act's detailed and expensive requirements for obtaining a mining permit. Under SMCRA, regulations required twenty-five separate sets of information. Among the most significant of these: (1) Names, addresses and phone numbers of all owners of equitable property, leaseholders, resident agents, directors of principal shareholders of the mine and contiguous areas; (2) Maps indicating the boundaries of the permit area, contour elevations, cross sections of pits, buildings, siltation ponds, storage piles, roads, and property ownership; (3) Climatological and air quality data; (4) A fugitive dust control program and monitoring plan for ambient air quality; (5) On and off site hydrological data on sources and quality of water, a plan for restoration and continued mining impacts on surrounding ground water resources; (6) A plan for revegetation including the species to be planted and irrigation methods.

A coal firm must submit and have approved each of these and the remaining nineteen requirements before it can disturb a cubic foot of earth. The costs of compliance with this permitting process alone can prove prohibitive for small firms or those with tight cash flows. Generating the information can take up to a year, a time horizon which smaller firms with limited resources can ill afford.

James McGlothlin, president of United Coal and a director of the National Independent Coal Operator's Association, contended: "The bureaucracy of acquiring a permit will totally eliminate most small . . . operators. . . . Most of these operators do not have registered or certified engineers, hydrologists, the equipment or the technicians to do core drilling, or most of the expertise required to obtain a permit under the Act and to continue compliance therewith" (Senate Hearings, Subcommittee on Lands and Natural Resources, 1977c, p. 303). Large firms, on the other hand, often have in-house engineering and the scientific knowhow necessary for compliance.

Through performance standards and information requirements such as those imposed by SMCRA, environmental regulation "transcends" the market in a way not envisioned during earlier regulatory regimes. Unquestionably environmentalists were aware of the anticompetitive effects of costly regulation. Legislators and coal firms raised this point consistently during the SMCRA debates, and in fact succeeded in providing for limited federal assistance to small coal firms. Nevertheless, the main emphasis in SMCRA remained stringent environmental standards, even at the expense of economic competition or lower prices.

New Actors

Attempts to enact and implement the new social regulation necessarily involved new actors in regulatory politics. Obviously environmentalists and other public lobbyists, the proponents of the new ideas, have added a new dimension to regulatory affairs. Other new actors, though, have appeared on the scene as well. In particular, individual business firms have become participants in regulatory politics. The new regulation has evoked firm-level participation often at the expense of traditional trade association activities. Moreover, environmental scientists and engineers have been brought to the fore within firms as managers attempted to cope with the new regulatory milieu. This professional shift mirrors a similar one in regulatory agencies. It also constitutes an important firm-level decision with profound implications for regulatory politics and policy.

Environmentalists exhibit decided differences with activist political elites of the Progressive Era and the New Deal. Perhaps the most noteworthy distinction between environmentalists and Progressives or New Dealers is the former's accent on participatory democracy. According to Carl Bagge, president of the National Coal Association and a longtime participant in regulatory affairs: "Politically, the consumer movement, the environmental ethic, the drive toward participatory democracy . . . are all pushing regulatory agencies to adopt policies that were regarded as irrelevant a few years ago" (1975, p. 187). Interestingly, Bagge was a key participant in the enactment of SMCRA. His primary concern, therefore, lay in the area of environmental regulation. Nevertheless, he recognized the political kinship between environmentalism and consumerism. In his view, the notion of participatory democracy was the hallmark of the two movements.

John Gardner, former president of Common Cause, described the emergence of this concept as follows:

> Then in the 1960s a feeling of citizen action reappeared with extraordinary vigor. . . . The peace movement, the conservation movement, the family planning movement emerged as potent elements in our national life.
>
> Much of this activity was diffuse, erratic and poorly organized. But little by little citizen action began to develop a more professional cutting edge. (1972, pp. 73–74)

Motivated by the concept of participatory democracy, environmentalists have carved out a role in the policy process. They have been in the forefront of Gardner's professional cutting edge. The most obvious example of this is the Nader organization, Public Citizen. The activities of Ralph Nader on behalf of consumers, beginning with the publication of *Unsafe at Any Speed* and continuing through a highly critical study of consumer protection at the Federal Trade Commission, are well known. Less known, but just as influential, were two critiques of environmental policy, *Our Vanishing Air* and *Water Wasteland*, both prepared by "Nader's Raiders" (Marcus, 1980). These reports not only reflect new ideas about government-business relations, but also recommend specific regulatory institutions and policies.

In his analysis of public interest groups, Jeffrey Berry concluded: "these organizations are slowly changing the overall environment within which governmental officials formulate public policy . . . public interest groups have been *consistent and enduring actors*, aggressively trying to influence governmental decisionmakers. . . . The opinion they can arouse, the bad publicity they can generate, the lawsuits they can file are all factors relevant to the deliberations of those who must make policy decisions" (1977, p. 289, emphasis added). Strategies and actions based on the ideal of participatory democracy, then, set environmental reformers apart as new regulatory actors.

Characteristically, national environmental groups such as the Sierra Club, Friends of the Earth, Environmental Policy Center and the Environmental Defense Fund vigorously supported SMCRA in its legislative phase and continued to monitor surface mine regulation during implementation. Their participation ranged from providing expert testimony at congressional hearings and performing technical studies on surface mining effects to seeking injunctive relief in federal courts against the leasing of additional public coal land while SMCRA was under consideration. Smaller environmentalist groups, many locally based, joined in the struggle to pass SMCRA by depicting, on a very personal level, the social and environmental impacts of coal surface mining. Such organizations as Virginia Citizens for Better Reclamation and the Northern Great Plains Resources Council regularly participated in hearings. In addition, SMCRA's provisions for a public role in the permitting process assured environmentalists a role in the implementation of SMCRA.

While environmentalism has brought public lobbyists to prominence, it also has served to undermine the role of private lobby groups. More specifically, environmental regulation has led to the emergence of a new kind of firm, staffed with new kinds of managers and professionals. Commenting on this, Peter Drucker noted: "top executives find that they have to take the initiative in areas of public policy. They have to identify issues, think through and develop concepts, policies and rules, and then have to take the lead in educating lawmakers, bureaucrats, their own industry and public opinion" (*Wall Street Journal*, June 17, 1978). Traditionally, these responsibilities had fallen to trade associations. The new state of affairs amounts to an increasing politicization of management.

Daily dealings with environmental regulation have accentuated the individual firm as a participant and exacerbated the collective action problems confronting business lobbies. This has occurred primarily as a result of imposing uniform performance standards. Such standards do not fall evenly on all firms for a variety of reasons. First, the individual profit functions confronting firms differentiate the impacts of environmental standards. Indeed some firms may be able to aggrandize their market position at the expense of competitors because they are better prepared to adjust to the new social regulation. This seems most obvious for big firms versus small firms.

Some corporations that can more easily afford to comply with proposed environmental regulations may tacitly support such regulations in the hope of expanding their markets at the expense of smaller firms with less financial latitude. Also, many large firms perceive regulation, once it is on the books, as reducing uncertainty in this era of public lobbies and participatory democracy. Therefore they may be favorably inclined toward regulatory initiatives, provided the benefits of lower risk are not outweighed by the costs of compliance. The reduction of uncertainty through increased regulation, though, is an option only for extremely large firms. For smaller firms, increased federal regulation may also increase certainty by putting them out of business.

This entire situation has led environmentalists into a working relationship with larger corporations without regard to the impact on the market. It should be noted, however, that environmentalists do not ordinarily form a Machiavellian alliance to drive smaller firms out of business. The situation is more subtle. In addition to a loose mutual interest in passing regulatory legislation, the environmentalists and the larger corporations are relatively comfortable with one another in that they have comparable levels of technical expertise and political sophistication.

This can lead to closer communication if not a more "amicable" relationship. Larger firms often have environmental engineers or governmental affairs personnel on their staffs. This common ground comes out clearly in congressional hearings, which often turn out to be dialogues between environmentalists and legal-technical representatives of big business. The obvious ideological rift between them over capitalism remains hidden under a veneer of technical language and political savvy. Yet cooperation is facilitated at the expense of smaller firms.

While the interests of large and small firms seem easily separable under the imposition of uniform performance standards, large firms may differ among themselves too. The economic-financial situation confronting large individual firms varies; and these variations can affect rational firm responses to environmental regulation. It is important, in this respect, to recognize that earlier price and entry regulation, focusing on an-industry-at-a-time (and usually smaller, more cohesive industries at that), facilitated the development of a common industry position and political leadership by business lobby groups. Environmentalism and other forms of the new social regulation, on the other hand, encompass many firms in many industries, thereby presenting aggregate business actors with severe collective action problems.

Yet another factor differentiating the impact of environmental performance standards on firms is our federal political system. As Daniel Elazar notes, one virtue of federalism is a measure of political flexibility or experimentation (Elazar, 1966). With respect to environmental regulation, federalism has fostered considerable variation among the states in terms of scope and stringency in performance standards. Thus individual firms, depending on the location of their operations, must comply with more or less tough environmental regulations. Obviously those already in compliance with relatively strict standards may enjoy an advantage with the imposition of uniform national standards. This impact of federalism further highlights the participation of individual firms and erodes the effectiveness of trade associations.

Interestingly, trade associations also represented new actors in regulatory affairs. In part, the older associations changed in adapting to the new regulatory regime, and in part the regime brought forth completely new associations. The variety of trade associations reflected differences among coal firms with respect to SMCRA.

The American Mining Congress (AMC) and the National Coal Association (NCA) figured prominently in SMCRA during its legislative and implementation phases. These are national-level organizations. AMC represents all major mining industries in the United States, including copper, sand and gravel, gypsum, iron ore, uranium and other hardrock minerals in addition to coal. Representing such diverse interests, it is difficult for AMC to work out a strong political position unless there is widespread support (i.e., homogeneous preferences) among its membership. NCA represents large and many midsized coal firms. NCA, while smaller and considerably less diverse than AMC, still faces a serious collective action problem given the significant differences among coal mining firms.

It is important to recognize that one impact of the new regulatory regime that paralleled the politicization of individual firms was the increased capability of trade associations. According to Raymond Peck of NCA, "There was no independent [i.e., NCA] input into the 1975 bill and the Administration developed its veto position pretty much on its own" (personal interview, May 11, 1980). In fact, Peck's move from the Ford Commerce Department to NCA symbolizes the effort to increase capability. In Peck's words, "the 1970s were a period of learning, reaction and development by the coal industry" (ibid.). Prior to SMCRA's introduction, all coal firms had little to do with the Washington bureaucracy, even through trade associations. If SMCRA did not make NCA a new actor, it at least forced it to play a new role.

Two other national-level trade associations involved with SMCRA were the National Independent Coal Operators Association (NICOA) and the Mining and Reclamation Council (MARC). NICOA, though a national organization, in fact represented eastern coal interests. Its smaller midwestern members from Ohio and Indiana did not differ significantly from eastern members. Thus, NICOA enjoyed a much more unified membership than either AMC or NCA, though it too had to adapt to the new social regulation.

MARC was a genuinely new actor. It did not come on the scene until 1977. Founded by Ben Lusk, former president of the West Virginia Mining and Reclamation Council, MARC was an explicit response to SMCRA. According to Dan Jerkins, vice-president of MARC, this trade association seeks "expansion of the coal industry through environmentally sound practices" (personal interview, February 2, 1981).

MARC represents primarily surface mining firms. Its membership includes five hundred firms (compared to roughly seventy for NCA) ranging in size from 100,000 tons per year upward to huge western firms. Also, its membership covers thirty-five states and it has attracted a number of newer western firms. Its membership does not include the older interregionals, the mainstays of NCA.

Finally, various state trade associations began to take part in the SMCRA debates. Their roles, for the most part, fit the regionally based interests of smaller and medium-sized coal firms. After SMCRA's passage, these trade associations either ceased political activity or shifted to legal action. For example, the Virginia and Indiana surface mining associations sought to overturn SMCRA in the federal courts. Again, at a minimum, this represents a new activity for these organizations.

Another important new actor in regulatory politics is the congressional staff member. Congressional staff emerged as important policy actors concerned with the new social regulation (Malbin, 1980). It is impossible to overstate their role in abetting and, at times, orchestrating the participation of environmentalists during the legislative phase. After SMCRA's passage, the staffers who moved on to organize and administer the Office of Surface Mining Reclamation and Enforcement (OSM) provided environmentalists with continued access to policy-making machinery in the federal bureaucracy. The former staffers and environmentalists they selected to assist at OSM helped to make the agency a new actor, one more responsible to public lobby concerns than to business advice, one that actively and aggressively sought input from environmentalists in promulgating regulations.

Two examples of the important role of congressional staffers are Norman Williams and Don Crane. Williams came from the West Virginia state agency responsible for surface mine regulation (West Virginia has had a surface mining law since 1939) to serve on the staffs of both Patsy Mink and Morris Udall. After working closely on SMCRA, he became an assistant director of OSM. Crane, another alumnus of Udall's staff responsible for SMCRA, was appointed regional director of OSM's western operations. The important point is not that these two fashioned careers for themselves. Rather, the point is that as environmentalist partisans and authors of SMCRA, they brought with them to OSM a keen interest in effectively implementing the law.

The fact that officials such as Williams and Crane administered SMCRA facilitated environmentalists' efforts to write detailed regulations to restrict firms' latitude in compliance. Detailed regulations emerged, despite the coal industry's objections, suggesting a measure of success in establishing a new subgovernment.

The final set of new actors in the political-economic framework is the federal bureaucracy. Any firm, in deciding on a behavioral course, must take account of the relevant executive agencies and cabinet departments. Regarding SMCRA, these included the Federal Energy Administration (FEA), the Department of Energy (DOE), the Council on Environmental Quality (CEQ), the Environmental Protection Agency (EPA), and most important the Department of Interior (DOI). In the 1970s all of these agencies were either new or called upon to play new roles.

Although the two energy agencies, FEA and subsequently DOE, were not part of SMCRA's earliest legislative history (they did not exist), they played a prominent role during the 93rd, 94th and 95th Congresses. As intimated previously, after 1973 national policy concerns with adequate energy supplies figured in the enactment of SMCRA. Logically, therefore, these two agencies were major actors. This role was enhanced considerably when responsibilities for coal mining data shifted from DOI to DOE.

The passage of S.425 and President Ford's two vetoes of that bill dominated SMCRA's legislative history during the 93rd and 94th Congresses. At that time FEA, under the direction of Frank Zarb, led the fight on Capitol Hill to justify the Administration position. As spokesman for the Administration, Zarb asserted that SMCRA would: (1) put up to 36,000 people out of work; (2) reduce coal production by 40–162 million tons/year; (3) significantly increase utility costs; (4) increase our dependence on foreign oil. FEA staunchly defended these conclusions before House and Senate subcommittees in the 94th Congress and was instrumental in putting together an Administration version of SMCRA much more favorable to the coal industry in general.

After the 1976 general election, FEA and DOE made a 180-degree shift in policy, as Jimmy Carter had pledged to sign a federal surface mining law. In contrast to Frank Zarb, John O'Leary, the new FEA director, told a Senate Subcommittee on Public Lands and Resources that "the need for minimum national performance standards is further illustrated by recent State actions . . . competitive pressures among States have . . . resulted in failure to enact more stringent laws and have often led to less vigorous enforcement" (Senate Hearings, 1977c).

The DOE occupied an extremely awkward position with regard to SMCRA. Created under the Energy Reorganization Act of 1977, DOE took as one of its primary policy objectives the increased utilization of coal to generate steam electricity. DOE always accepted the concept of SMCRA in principle. However, in the implementation phase, the Department took a position that OSM was overly aggressive and overly restrictive in the specific regulations it devised. DOE along with economists from President Carter's regulatory review group argued that OSM ought to set performance standards on environmental quality alone, leaving it to coal firms to meet those standards (*Wall Street Journal*, January 2, 1979). In this way, it was hoped, environmental regulation could be made cost effective and coal production could increase still faster.

The CEQ and EPA logically played very important roles in the legislative history of SMCRA. Each agency was itself a product of the environmental movement and provided expert advice and assistance in drafting the legislation. The CEQ, an advisory agency under White House control, helped to draft a surface mining bill under the Nixon Administration (S.993) in which it gave qualified support to SMCRA. Some of the economic information in this report, though, was rejected by President Ford when he formulated his veto message. CEQ officials did manage to get this information into the hands of pro-SMCRA legislators to contest the Ford figures on unemployment and coal production losses. On balance, CEQ and its technical-oriented staff supported SMCRA within the parameters of its political abilities and sometimes beyond as indicated by the leak of information deleterious to Ford's veto strategy.

The role of EPA in SMCRA was at once more direct and more complicated than that of the CEQ. "The EPA became involved in surface mining largely through its responsibility to regulate effluent discharges," according to William Dickerson, director of EPA's Office of Environmental Review (OER) (personal interview, July 21, 1980). OER served as a legislative activities center for the agency, thereby placing Dickerson directly in SMCRA debates. Under SMCRA, OSM was to regulate effluent discharges from coal mines. This gave rise to the familiar problem of overlapping jurisdiction within the federal bureaucracy. Dickerson pointed out that EPA kept a "low profile" through the legislative and implementation phases, working out memoranda-of-understanding with OSM. Nevertheless, a real potential for a "turf war" between EPA and OSM developed over water pollution control responsibility.

The emerging relationship between EPA and OSM was complicated by the former's diffidence in regulating what is known as "nonpoint source" water pollution (NPS). Point sources, or direct effluent discharges from a plant, were clear enough. EPA's attempts to regulate nonpoint sources though, had raised a number of legal and definitional problems for the agency. Consequently, EPA had been steering away from the problem. SMCRA, however, raised the issue. Acid runoff into a stream from a mine miles away constitutes nonpoint source pollution. OSM and pro-SMCRA forces, moreover, adamantly insisted on regulating such water pollution. Thus, questions arose over both jurisdiction and the proper emphasis for EPA's control of effluent discharge programs.

In light of these circumstances, Joseph Krevac, director of water criteria and standards at EPA, "made a conscious decision to leave nonpoint problems to OSM" (personal interview, July 22, 1980). In Krevac's view, EPA had matured and developed a sense of responsibility which implied a sensitivity to the effect on businesses of ambiguous or overly specific regulations. Dan Deely, having worked at EPA on NPS programs and on SMCRA, offered a different interpretation. He accused EPA of "pussyfooting on definitions" with NPS. He further pointed out that in 1974, the "208" regulations on water planning programs authorized EPA to veto pollution control plans of individual plants on the basis of NPS considerations. In his view, EPA had abandoned this responsibility and became extremely uncomfortable when OSM dredged it up (personal interview, July 22, 1980). This jibed with William Dickerson's observation that occasionally firms approached EPA for support against OSM's position on a particular regulation: over a period of years, business and EPA had both moderated their positions toward one another.

In sum, EPA played a somewhat ambivalent role in SMCRA. On the one hand, like CEQ it offered technical assistance and, beyond that, helped to write certain provisions and regulations. On the other hand, the hydrological aspects of SMCRA made EPA and OSM direct rivals. More significantly, the tough approach of pro-SMCRA people and OSM flew in the face of EPA's efforts, in the mid-1970s, to apply performance standards more flexibly and to avoid getting into regulation of production processes themselves. Director Krevac viewed such activities necessitated by NPS as "a potential beehive." OSM exhibited no such inhibitions.

The final and most important executive branch actor in the political-economic framework was DOI. Within DOI, the Geological Survey (GS), Bureau of Mines (BM), Bureau of Land Management (BLM), and OSM were intimately involved in SMCRA. The GS played as close to a completely neutral role as possible. It simply provided data as requested by Congress and other bureaucratic actors. The BM not only provided data on the number or nature of mines and mining operations, it also performed analyses to determine the likely economic impacts of SMCRA on "typical" surface mines in various regions of the country. Both the GS and BM played insignificant roles in SMCRA after the 93rd Congress.

The BLM, however, was involved all along. At first, it served a similar informational function to that of GS and BM. Later that changed as the BLM received responsibility in 1975 for administering the so-called "211 regulations," which empowered it to oversee surface mining and reclamation on public lands. Coal firms lease public lands for mining through BLM. The "211 regulations," promulgated under the Ford Administration, set guidelines for mining and reclamation on these lands and vested the regulatory authority in BLM. Subsequently, spokesmen for the Ford Administration pointed to the "211" program and newer state surface mining laws as a rationale for vetoing s.425. Wittingly or unwittingly, then, BLM served to make SMCRA unnecessary in the view of some. Similarly to the FEA, though, this role changed after the 1976 election and the "211 regs" were seen as bolstering SMCRA but not providing a substitute since they did not cover eastern and midwestern coal mining on private lands. BLM did provide valuable experience for western firms in complying with the "211 regs."

It is important to note that BLM and its sister bureaus in DOI were generally guided by the philosophical outlook at Interior. Until the 1970s there was no question that DOI's main mission was to facilitate the development and exploitation of America's natural resources. The generation of data, research on mining or exploration techniques and leasing facilities all served this purpose. Initially, this pervasive attitude impelled environmentalists to argue for giving EPA responsibility for SMCRA (Matthes, 1977). However, the obvious accommodationist stance of EPA in the mid-1970s, what Krevac called "maturation," soured them on that idea. The solution was to create a new bureau within DOI that would be controlled by environmentalists and their congressional allies. In addition, environmentalists generally sought broader influence in Interior and, to a degree, achieved it as exemplified by the "211 regs." Of course, Carter's election and the philosophical bent of Cecil Andrus greatly abetted this process.

New Institutions

If the ideas justifying government regulation of business have changed, so too have the institutions supported by those ideas. Institutions should be construed to encompass both actual government organizations and the legal-administrative processes under which they act. In speaking of a qualitative shift in regulation or a new regime of government-business relations, institutional changes are readily discernible and clearly reflect the new policies advocated by political movements.

For their part, Progressives strove to construct institutions that would be insulated from politics. This would be accomplished, or so reformers thought, by devising independent regulatory commissions. According to a congressional research staff report, in the late nineteenth and early twentieth centuries "faith in experts and distrust of politics seem to be the major historical motivations for the independent status of the regulatory commissions" (Senate Document 95-91, 1978). These hopes proved ill-founded. Regulation could not be insulated from politics because it imposed cost and bestowed benefits on different groups in society. Moreover, operating in a climate made hostile by a basic societal suspicion of federal intervention, independent commissions were constantly aware of the need for keeping a low profile. As a result, the commissions tended to operate as something less than the vigilant guardians of public welfare envisioned by the Progressives.

New Deal reformers recognized this flaw in regulatory commissions. They responded by attempting to institute presidential, or executive, government; to bring federal regulation directly under the control of the White House (Milkis, 1981). If the commissions constituted a fourth branch of government, largely under the sway of the regulated industries themselves, the solution was to establish executive authority to insure that the public interest was served. This approach proved unsatisfactory for two reasons.

First, although the New Dealers firmly controlled the executive branch, in the 1930s the Congress and the political parties were still quite resilient. This meant that the New Dealers could not realize their goal of presidential control without hard bargaining and compromise. Without thorough executive control, the same problems that bedeviled the Progressives cropped up. In addition, the New Dealers appear to have overlooked a crucial reason for the failure of commissions: they were organized on an industry-by-industry basis. This can be seen in the New Deal agencies such as the Federal Communications Commission (FCC), Civil Aeronautics Board (CAB) and Securities and Exchange Commission (SEC) as well as their predecessor, the Interstate Commerce Commission (ICC). Even agencies which made broader efforts to regulate business such as the Federal Trade Commission approached their tasks on a case-by-case basis, which amounted to the same thing.

In this traditional mode, regulatory policy dealt with questions of price, entry and business practices in a particular industry. Regulators dealt, on a day-to-day basis, primarily with industry people. This situation facilitated the development of a close relationship between industry and agency. It also fostered the so-called "revolving door" whereby regulators could move into private industry in time. This classical argument was developed by Marver Bernstein in 1955: "The objective is to make the survival of the agency dependent upon a working alliance with the regulated groups on terms dictated largely by those groups. . . . For regulated groups the regulatory process may be one method of converting public power to private gain" (Bernstein, 1955, p. 266).

The co-optation of commissioners took place in stages, prompting Bernstein to talk of a life cycle of commissions in which they changed from vigorous and aggressive agencies to relatively complacent and hidebound organizations. Though the imagery of industries "capturing" agencies is overdrawn, Bernstein does have a point.

The second reason the New Dealers' scheme of presidential government failed ultimately is that they could provide no electoral guarantee that the occupant of the White House would be sympathetic to federal regulation. This became apparent to environmentalists and consumer advocates under the Nixon Administration. Like FDR, Nixon wanted to control the bureaucracy. However, his vision of Washington's proper regulatory role was quite different from Roosevelt's.

Surveying the attempts of Progressives to remove regulation from political arenas and New Dealers to impose presidential control over regulation, contemporary activists have struck out on a new path. They have learned from past "mistakes." Like New Dealers, they recognize that regulation is inherently political. They have tried to carve out for themselves permanent institutional niches in regulatory arenas.

Environmentalists have recognized that passing a bill is only the beginning of regulatory political processes that could well emasculate an agency in the long run. The permeability of the American political system requires an ongoing effort to represent a public interest in policy forums. Cognizant of this, environmentalists have sought the appointment, in key policy or administrative posts, of individuals openly sympathetic to their cause. This is a clear departure from seeking impartial experts as administrators or presidential control or regulation. As Richard C. Leone points out, "With regard to administrative regulation generally, public interest advocates . . . have one thing in common: they all seek the reform and reorientation of regulatory agencies" (1976, p. 48). This has been a deliberate effort to change the institutional framework of government-business relations by bringing to bear a new set of ideas about public policy and democratic politics.

Alfred Marcus demonstrates this in his analysis of the EPA. He shows not only that the agency and its enabling legislation were "informed by a theory of how to best prevent a regulatory agency from being 'captured' by an industry or afflicted with bureaucratic sloth," but also that this theory (interest group liberalism) figured in the creation of the EPA and enactment of environmental laws because of "the work of middlemen who transmitted the notion to politicians" (Marcus, 1980, pp. 267–70). These so-called middlemen were environmentalist leaders operating through their public lobby organizations.

Institutions in the contemporary regulatory regime have changed also in terms of the legislation empowering federal agencies and bureaus. This change follows logically from environmentalists forsaking the ideas of neutral, nonpolitical agencies and a free market. As long as regulatory bodies were taken to be nonpartisan, it was permissible, even desirable, for Congress to bestow a broad mandate on bureaucratic experts who would, in turn, generate the specific rules necessary for implementation. However, once environmentalists and their supporters became convinced that a more-than-congenial relationship had developed between agencies and industry, their preference naturally turned to specific legislation and detailed performance standards.

In their study of air pollution regulation, Ackerman and Hassler discovered just such a self-conscious environmentalist strategy: "Operating under a vague statutory mandate, New Deal agencies have ample opportunity to fend off statutory interventions. . . . In contrast, the Clean Air Act had tried to resolve so many disputable substantive issues in 1970 that recurrent Congressional reconsideration was fundamental to the policy-making process" (1981, p. 26).

In addition to detailing rules and responsibilities for regulators in enabling legislation, environmentalists saw another remedy to the weakness of regulatory institutions. The agency-industry relationship could be moderated by restructuring regulatory policy to cut across a number of firms, industries and even economic sectors. Remarking on this institutional approach, Paul McAvoy observed: "In its day-to-day operations to regulate both air and water quality, EPA turned to setting limits on pollution emitted by a plant or vehicle per unit of operation—what might be termed product performance standards—even though the Clean Air Act was framed in terms of regional goals . . ." (1979b, p. 84). Moreover, responsibility for regulation need not be lodged in a particular agency. Previously an industry's or firm's contact with government was limited, for the most part, to dealing with one agency or department on a particular issue or set of issues. Now businesses in the normal course of operations confront a wide array of regulatory bodies administering a variety of programs across a spectrum of business interests. Under this new rubric it is difficult to establish the traditional kind of relationship that had existed between industry and agency because regulation is organized on a functional rather than an industry basis.

The specific institutions changed or created under SMCRA represent an application of the lessons learned from previous regulatory experience. Environmental lobbies, new institutions in their own right, sought to structure SMCRA to accommodate ongoing participation by public interest groups in the implementation and administration of the law. In this regard, SMCRA provides for public hearings on permit applications. Under SMCRA, an applicant for a surface mining permit must publicize his application, replete with all technical information, in a local newspaper once a week for four weeks prior to a public hearing. Moreover, "Any person having an interest which is or may be adversely affected," may file an objection to the application and proceed with informal review, formal regulatory hearings and appeals (30 USCS, sec. 1263).

In addition, the regulations provide that private citizens may request spot inspections of coal mine operations by regulatory officials if the citizen can give the official "reason to believe that a violation exists" (30 Code of Federal Regulations [CFR], 842.12). These provisions afford environmentalists an institutionalized means of monitoring coal surface mining.

SMCRA also endows them with automatic standing in citizen suits against coal firms or the regulatory authority. In this way, they can supplement the activities of inspection and enforcement personnel authorized by the Act. More important, this new institutional arrangement can serve as a prophylactic measure to guard against the close industry-agency relationship found under earlier regulation.

OSM itself was intended to be a distinctly political agency, staffed by ecology-conscious partisans who would be impervious to attempts at "capture" by the coal industry. To insure this, SMCRA provides that OSM may employ people from DOI or other federal agencies only "providing that no legal authority, program or function in any Federal agency which has as its purpose promoting the use or development of coal . . . is so employed" (30 CFR, sec. 1211). With such people in office at OSM, environmentalists could forge a subgovernment in which they, rather than business firms, would be the insiders.

Last, SMCRA reflects the new regulatory regime's institutions insofar as the program is designed and administered on a functional rather than an industry basis. Superficially, this may seem somewhat anomalous since the program pertains to a single industry, coal mining. However, as originally conceived, SMCRA encompassed all surface mining. This is suggested by the designation Surface Mining Control and Reclamation Act. The law, in its present form, in fact pertains to the "surface impacts incident to underground coal mining" as well as surface mining per se (Jones, 1979b, p. 2). In effect, then, the law cuts across two coal mining industries, each with its own technology, markets and financial base. That SMCRA was devised along functional lines is evident from the fact that Title VIII directs OSM to study the feasibility of extending SMCRA to other extractive industries utilizing surface technique; for example, sand and gravel, iron ore or copper. As we shall see, certain political realities focused SMCRA on coal initially. Nonetheless, SMCRA, in its functional organization, fits the institutional template for environmental regulation.

New Tasks

Obviously, the success of environmentalism and the consumer movement has set new tasks for the federal government to perform. The substance as well as the institutional structure of environmental regulation accounts for these new tasks.

Under the regulatory regimes of the Progressives and the New Deal, federal agencies perform mostly adjudicatory functions. This was natural for regulatory commissions charged with ruling on prices and entry. The professions that dominated regulatory affairs through the mid 1960s were law and, to a lesser extent, economics. Adjudication of price and entry controversies took place through the highly technical institutions of administrative law. It demanded evidence and expertise that only economists and lawyers could provide.

Environmental regulation is quite another matter. It entails tasks of a more legislative and executive than a judicial nature. Environmental regulation, by conscious design, calls for the development of uniform national performance standards for all businesses. Congress has called upon regulators to write these detailed performance rules.

In addition to this technical rulemaking responsibility, environmental regulation requires that the federal government develop environmental impact statements (EISs) to insure that proposed business projects do not cause undue harm to the surrounding ecology. These two kinds of tasks demand a new kind of expertise, that of environmental scientists and engineers. The language and research involved in regulation has shifted from law and economics to biology and engineering.

Lawyers are still involved, but there is also a new breed of lawyers, activists who see themselves as serving public interests rather than business organizations. As the environmental movement desired, these ecology-conscious lawyers have moved from citizen lobby groups into federal regulatory positions. We shall see that they often view themselves as law enforcement officials, sworn to uphold the regulations of their agency. They frequently carry out surprise inspections, sometimes even arriving at a firm's operations incognito. This kind of executive activity is unprecedented. It reflects both the ideology and regulatory institutions of environmentalism.

By creating a new agency, the Act established new tasks, both for OSM's staff and the Secretary of Interior, in whose department Congress lodged OSM. The secretary, for example, must "review and approve or disapprove State programs for controlling surface mining operations and reclaiming abandoned mined land" (30 USCS, sec. 1211). A variety of related tasks, including the promulgation of rules and regulations to deal with inspection and enforcement, fell to the secretary and OSM. Under SMCRA, a full staff of inspectors under the control of regional directors shouldered the responsibility of implementing the law. The regulations specify how frequently inspections must take place and what the penalties available for enforcement are (30 CFR, 843). In the 141 pages of the Act and 431 pages of implementing regulations, the surface mining and reclamation program also calls for a variety of technical expertise and scientific procedures not ordinarily found in the federal government. SMCRA requires the government, through OSM, to develop and analyze new kinds of information. Besides OSM, existing agencies including the GS, BM and EPA help to fulfill these tasks.

An Illustration of the New Regulatory Politics

SMCRA, it has been argued, reflects environmental regulation and, therefore, the new regulatory regime. The fundamental ideas that distinguish the new regime are embedded in the law and its companion regulations. These ideas, furthermore, have shaped the institutions, tasks and actors—the outward manifestations of the new regime.

At this juncture it may be useful to take a brief look at how the political-economic framework and subgovernment confronting coal firms operated. Fortunately, there is a convenient example from the early stages of the fight to enact SMCRA. The decision to regulate only coal surface mining shows clearly how the new social regulation has changed regulatory politics. Significantly, it also shows how traditional interests and elements of American politics have meshed with the new ideas, actors, institutions and tasks embodied in SMCRA.

As originally conceived, SMCRA was to regulate all surface mining in the United States. In its final form, the only reference to noncoal surface mining was a title of the Act allocating funds for studying the possibility of regulating surface mining in other mineral industries. How did this version of SMCRA come to pass? Answering this question not only highlights the role of congressional politics in environmental regulation, but also sets out the cast of less visible, though crucial, actors confronting coal firms.

One argument for regulating the surface mining of coal separately hinges on mining law. Because coal is distinguished from so-called hard-rock minerals in the U.S. Code, it may seem logical to perpetuate that distinction. Under the Coal Leasing Act of 1928, a company finding coal must lease the land, public or private, on which that coal is situated. Hard-rock minerals, though, remain subject to the Mining Act of 1872, which provides that anyone finding a mineral deposit on public land may claim it outright. The fact is that a "leasable" resource, coal, is distinguished from a "locatable" resource, hard-rock minerals. On the basis of this precedent, there may be a presumption in favor of treating the two separately in other regulatory legislation such as SMCRA.

However, this approach would be at variance with the new regime, which sought to move away from the old formula of industry-based regulation and toward a functional organization. The original bill to regulate all surface mining, indeed, proposed a functional solution. The key question, then, is not why regulate coal separately, but rather, why did environmental forces accept a regulatory approach they knew to be fraught with prospects for collusion between coal firms and regulators?

According to Norman Williams of OSM, he and other environmentalists had to recognize that "Coal is a different critter" (personal interview, July 28, 1980). Besides being treated separately in mining law, coal differs from other surface-mined minerals in that it is bedded in seams, constitutes an energy resource and causes severe off-site pollution problems (i.e., stream pollution and acid rain). Consequently, environmentalists had to treat coal separately. Coal surface mining raised a number of unique technical and engineering problems for regulators.

While this argument certainly has merit, it does not quite "ring true." It is the type of argument one would expect from a firm. To accept Williams' argument, one would have to believe that environmentalists put together an effort to regulate surface mining without carefully considering the operations of the various industries to be controlled. This assumption would be totally at variance with the preparation and expertise environmentalists demonstrated in all legislative battles. In addition, if they were that naive, it is difficult to believe that they would so uncharacteristically reverse themselves as quickly and quietly as they did simply because mining firms suggested it.

The only satisfactory explanation revolves around congressional politics. Offering his personal explanation, Raymond Peck of NCA stated that "In sum there were aggressive forces for excluding noncoal from stripping regulation, particularly copper interests, while there were aggressive forces for including coal" (personal interview, May 12, 1980).

Understandably, the hard-rock surface operators were foremost among the exclusionary forces. These included not only the copper interests Peck alluded to, but, just as important, sand and gravel miners. Both sets of interests, moreover, were in politically advantageous positions. As the chief counsel for an eastern-based interregional noted, both were "plugged into the federal government for a long time" (personal interview, October 30, 1980). The sand and gravel mining firms enjoyed considerable influence in Congress through their relationship with legislators concerned with the Federal Highway Trust Fund. This connection with a "sacred cow" gave them a powerful voice in any pending legislation. In addition, the sand and gravel industry had a direct voice through the AMC, which openly opposed SMCRA.

AMC also supported copper interests in an effort to exclude hardrock minerals. This is hardly surprising since hard-rock firms dominate AMC. Even though large coal firms belong and derived some sympathy from hard-rock firms, AMC, as any trade association, was governed by the most signficant view among its members. In this case, that view was to exclude noncoal. In early congressional hearings on SMCRA, the view was espoused consistently by AMC, National Limestone Institute, National Crushed Stone Association, Phosphate Lands Conference, American Iron Ore Association, National Land and Gravel Association and National Industrial Sand Association.

From all indications, this was a purely political view. Trade association people, coal firm executives, environmentalists and congressional staffers alike pointed to the negative public image of the coal industry as a major influence on the predisposition of Congress. This presented noncoal operators with a perfect strategy: fuel the public image of coal as despoilers of the environment. Indeed, as Dan Deely, a former EPA official, noted from his experience in developing SMCRA, other mining sectors "consciously sought to dissociate themselves from coal to preclude being tarred with the same brush" (personal interview, July 22, 1980).

After successfully deflecting attention from hard-rock mining, AMC cooperated fully with NCA in combating SMCRA. In fact, a joint AMC-NCA committee was established in 1974 to represent coal mining interests; but only after SMCRA was written to regulate coal alone.

This effort of hard-rock mining interests, and AMC generally, did not suffice alone. Environmentalists had powerful allies in Congress as well. However, these lines of alliance were weakened considerably on the particular issue of surface mine regulation. The weakening stemmed from the fact that SMCRA was developed in the House and Senate Committees on Interior and Insular Affairs. The leading congressmen and senators were Morris Udall (D-Ariz.), John Sieberling (D-Ohio), Patsy Mink (D-Hawaii), John Melcher (D-Mont.), Henry Jackson (D-Wash.) and Lee Metcalf (D-Mont.). In the House, Udall and Mink, having acceded by 1972 to the chairs of subcommittees on Energy and Environment and Minerals, Materials and Fuels, played key roles. In the

Senate, Jackson, as chairman of the full committee, and Metcalf, in control of the Subcommittee on Mines and Mining, were preeminent. All six legislators staunchly supported the environmental cause. However, except for Sieberling and Mink, all hailed from western states with significant hard-rock mineral industries: Arizona, Washington and Montana.

An important connection existed between Udall and the Arizona copper interests. Coal operators (primarily interstate firms with a high degree of political sophistication), NICOA and NCA complained bitterly in private interviews that Udall (to a lesser extent Jackson and Metcalf) built a national reputation on environmental issues while catering to hard-rock interests back home. In a rare outburst, one coal executive, while discussing this, suggested, "Someone ought to take Mo Udall out behind a barn and explain the facts of life" (personal interview, January 23, 1981). Dan Deely, having left EPA for a position at the Agency for International Development (AID), felt free to single out Udall and his relationship with Arizona copper interests as a critical factor in focusing SMCRA on coal (personal interview, July 22, 1980).

Apparently when it came to the "political facts of life," Mo Udall needed no instruction. The unsolicited comments from participants on both sides of the SMCRA debate linking Udall to copper interests provide a compelling partial explanation for the bill's eventual focus on coal. In fact, the entire constellation of political and economic forces favoring the exclusion presents a convincing explanation. However, it is only half the story of congressional politics and SMCRA's focus.

A curious alignment of environmentalists, the United Mine Workers (UMW), some eastern states and, again, western legislators fought aggressively for the regulation of coal surface mining. Each had a powerful, if completely divergent, interest in regulating coal surface mining.

Without a doubt, the environmentalists had the most obvious reasons for regulating coal. Environmental problems affiliated with surface mining first reached the federal level through the attention focused on the quality of life in Appalachia by Presidents Kennedy and Johnson. In 1967, somewhat ironically, Morris Udall's brother Stewart, then the Secretary of Interior, issued the results of a three-year study on surface mining, primarily in Appalachia where the results of contour mining had left a scarred and damaged environment. The Udall Report, with its environmentalist orientation, fit neatly with the concern over Appalachian quality of life; policymakers were well aware of the dire ecological results of surface mining coal in Appalachia from Harry Caudill's

1961 exposé, *Night Comes to the Cumberlands*. From the outset, then, federal concern with the impact of surface mining focused on Appalachia where surface mining means coal mining.

Beyond the simple fact that coal surface mining had damaged the Appalachian ecosystem so blatantly, environmentalists focused on coal for other reasons. As indicated by the strategy of hard-rock operators, coal had a bad public image in general. Because the earliest attempts to improve the environment aimed at reducing air pollution, coal, the major problem in Midwest and East Coast utilities, became an object of study and concern for environmental lobbies.

Finally, environmentalists received a powerful impetus to focus on coal when, around the time DOI released the Udall Report, it became economically and technologically feasible to surface mine the vast tracts of western coal on the Northern Great Plains. This opened huge new acreages of public land and open range to the hazards of surface mining. Environmental activists, their allies in Congress and even some concerned coal companies sought to preclude the possibility of a "second Appalachia" in the delicate ecology of several western states. On August 21, 1971, the *New York Times* reported: "On a scale far larger than anything seen in the East, where acreages totalling half the area of New Jersey have been *peeled off* for coal near enough to the surface to be strip mined, portions of six western states—Arizona, Colorado, Montana, New Mexico, North Dakota and Wyoming—face a topographical and environmental upheaval." Spurred by these various concerns, environmentalists threw their full weight behind legislation to regulate surface mining, with the accent on coal.

Two kinds of intersts drew certain eastern states into a loose alliance with environmental lobbyists on the need to federally regulate coal surface mining. As indicated in table 3.1, several states had passed surface mining laws before SMCRA's enactment. Most of these laws and amendments emerged after surface mining entered the national agenda in 1971. State legislatures sought to obviate the need for federal intervention.

From the point of view of Pennsylvania, West Virginia and Ohio, these state attempts failed because there were no centrally determined standards. In the words of Norman Kilpatrick, director of West Virginia's Surface Mining Research Library: "Numerous firms that operate only in States like Ohio, Illinois, Pennsylvania, and West Virginia (i.e., those with strong regulation) feel keenly the unfair competition of stripped coal from the States with weaker surface mine laws" (House

Table 3.1   States with surface mining laws showing the year of enactment and the year the law was most recently amended*

State	Enacted	Most recently amended
Alabama	1969	1975
Arkansas	1971	
California	1975	
Colorado	1973	1976
Florida	1971	1975
Georgia	1968	1976
Hawaii	1975	
Idaho	1971	
Illinois	1961	1976
Indiana	1967	1974
Iowa	1968	1976
Kansas	1968	1974
Kentucky	1967	1976
Louisiana	1976	
Maine	1973	1975
Maryland	1967	1976
Michigan	1970	1972
Minnesota	1971	
Missouri	1971	1975
Montana	1973	1975
New Mexico	1972	
New York	1974	1976
North Carolina	1971	
North Dakota	1969	1976
Ohio	1972	1974
Oklahoma	1971	1972
Oregon	1972	1975
Pennsylvania	1945	1976
South Carolina	1973	
South Dakota	1971	1976
Tennessee	1972	1975
Texas	1975	
Utah	1976	
Virginia	1950	1976
Washington	1970	
West Virginia	1939	1971
Wisconsin	1973	
Wyoming	1973	1976

*Material obtained from state offices.

Hearings, 1977b, v. ii, p. 190). In surface mine law as in taxation, interstate competition for investment and employment tends to restrain individual states in their attempts to control business operations.

In the eyes of eastern environmental activists too, this makes state-level regulation of coal stripping unworkable. Testifying before a House Subcommittee on Energy and the Environment, Mike Mullins of the Knott County, Kentucky, Citizens for Social and Economic Justice complained, for example: "The question we are addressing today is whether there is a need for a Federal strip mine bill. Since the State has been doing such a 'good job,' they feel like we don't need one. I think there is a great difference between passing laws and enforcing them. The State of Kentucky is one of the greatest law-passing States I know of. They are one of the worst I know of at enforcing these laws" (House Hearings, 1977b, v. i., p. 51). Convinced that state governments, by and large, would not enact, or if they did enact, would not enforce surface mine regulation, environmentalists turned to Washington, D.C., where they sought federal measures to coerce surface mine operators into reclaiming the land they had ravished and to strictly control future mining practices. This view dovetailed with the interests of some eastern states.

The alliance to regulate coal surface mining received additional support from the UMW and states in which deep coal mining contributed significantly to local economies. For example, Senator Walter Huddleston (D-Ky.) asserted: "In the two years following enactment of the Coal Mine Health and Safety Law [MSHA], deep mining in Kentucky decreased by over ten million tons per year—and surface mining increased by over 22 million tons per year. . . . Anyone who has seen the utter destruction of uncontrolled, unregulated, and unreclaimed strip mines cannot help but be profoundly disturbed by this trend" (House Hearings, 1973c, p. 23). Huddleston was arguing for SMCRA in order to offset MSHA's negative effect on eastern deep-mining. He wanted to saddle surface miners with their own federal regulation. This was precisely the strategy of the UMW in the early 1970s when they voted to support SMCRA.

Further, as Ray Peck of NCA indicated, most western coal workers were not unionized, a fact that induced the UMW to view SMCRA as a means of weakening the economic advantages of nonunion coal (personal interview, May 12, 1980). When the full impact of SMCRA's provisions on the surface effects of underground mining became apparent, the UMW rescinded its support.

The final element in this quadripartite alliance directed against coal surface mining was western legislators. Control of responsibility for SMCRA lay in the hands of Udall's Subcommittee on Energy and the Environment, Mink's Subcommittee on Minerals, Materials and Fuels and Metcalf's Subcommittee on Mines and Mining. As we know from Fenno's analysis of the Subcommittees on Interior and Insular Affairs, western legislators and numerous bills of local interest dominate those Committees — at least under earlier regulatory regimes. The congressional staffs of Udall, Mink and Metcalf independently and collectively steered these subcommittees toward SMCRA and broad environmental concerns. However, the local factors were not eradicated, only submerged.

Western legislators were plainly interested in developing the vast western coal reserves. This applies particularly to Udall and Metcalf. The various versions of SMCRA, all with strong emphasis on approximate original contour and steep slope provisions, were generally recognized as more severe on the East than the West. Moreover, western surface miners had been in compliance with tough state laws and reclamation programs administered by the BLM on publicly leased lands. Under these circumstances, federal regulation could prove advantageous to western coal especially after 1973 as the nation sought to reduce its dependence on oil.

Not surprisingly, the small eastern firms, those hurt the most by SMCRA, pointed this out. John L. Kilcullen, general counsel for NICOA, asserted that Udall, Melcher and Metcalf "wanted to shift the coal industry westward" and saw SMCRA as a means of facilitating this (personal interview, February 2, 1981). This is consistent with the local or regional bias traditional on Interior subcommittees reflected in their support for excluding hard-rock mining from SMCRA. Western governors supported SMCRA too, though with the understanding that states would retain primary regulatory authority.

The decision to regulate only coal under SMCRA illustrates a number of points about the political-economic framework, or subgovernment, in which the legislation was enacted. First, environmentalists showed a measure of political savvy. In light of the influential forces favoring the exclusion of noncoal surface operators and the fortuitous grouping of forces desiring regulation of coal surface mining, environmentalists worked for and obtained a reasonable compromise.

Since, for a variety of reasons, coal was their major concern, they agreed to regulated coal immediately while writing into SMCRA a title directing study of the problems associated with hard-rock surface mining. They also abandoned initial attempts to outlaw surface mining. Following on this conclusion, the hard-rock mining interests showed that a unified group with very specific goals and arguments, that would not clearly do violence to the most fundamental ecological concerns, can wring concessions from environmentalists and their allies.

A major reason for the political sophistication of environmentalists and the success of the hard-rock interests was the congressional staffs. Because the staffs, in this case of Udall, Mink and Metcalf, were almost solely responsible for researching, drafting and shepherding SMCRA through Congress, they were in a position to expertly advise their environmental allies. Moreover, it was in the interests of the staffs (career advancement) and their legislators (national reputation) to enact some strong piece of environmental legislation on surface mining; and they took the most they could get.

The decision to regulate coal also reteaches the old lesson about congressional politics that the legislators, especially in the House, will serve their constituents. Although this holds particularly for western legislators serving on Interior and Insular Affairs Committees, it is true in general, as illustrated by success of midwestern legislators in amending sulfur standards to benefit their region's coal firms. The nature of environmental regulation, especially SMCRA, which applies uniform standards to regionally diverse economic operations, evokes local constituency service responses.

Conclusion

Although the preceding elaboration of the political-economic framework in which SMCRA developed raises several problems and considerations for our analysis of firm behavior, it is, in fact, only a crude representation of the complexity facing individual firms. Whether each firm perceives the complexity or whether a rational firm need concern itself with that complexity is another matter. The fact remains that the political-economic framework, the actors, ideas and institutions confronting the individual firm, play a crucial role in the firm's determination of a behavioral course and the success or failure of any attempt to influence regulatory policy.

4 The Coal Industry and Firm Characteristics

Historically, the coal industry has resisted vigorously federal regulatory initiatives. Until the emergence of the new social regulation, the industry succeeded, for the most part, in avoiding interactions with Washington: "Until recently, coal was the least politicized of the nation's energy sectors. Aside from regulating mine safety, labor relations and private access to coal on federal lands, Washington largely left coal in the hands of the private sector" (Rosenbaum, 1978, p. 22). The imposition of mine health and safety regulations in 1969, the indirect impact of the Clean Air Act and the prospect of SMCRA, however, all served to politicize the coal industry in the 1970s. To be more precise, these measures brought the coal industry into federal regulatory politics. Coal firms and industry organizations had long been involved in legislative affairs at the state level, but had fought the federal government at every turn. Writing on the political history of coal mining, James P. Johnson observed: "The southern wing of the industry, allied with a variety of other operators, fought the NRA throughout its history and eventually had it declared unconstitutional" (1979, p. 240).

The industry's response to Roosevelt's NRA echoed its response to Woodrow Wilson's Fuel Administration. Only the emergency of World War I won grudging acceptance of the latter from coal firms, even though the federal intervention promised economic stabilization. As Johnson concluded, "The politics of soft coal was . . . a story of a splintered industry unable to find a means to save itself through industrial self-management" (ibid., p. 245).

We can draw two important inferences from this portrait of the coal industry. First, coal firms did not behave as economic theory predicted.

Not only did they not "demand" regulation as Stigler and his disciples would have predicted, but they rejected federal initiatives such as the Guffey Acts that sought to revitalize the coal industry. This behavior is all the more remarkable in light of Richard Posner's assertion that "the demand for regulation is greater among industries for which private cartelization is an unfeasible or costly alternative—industries that lack high concentration and other characteristics favorable to cartelization" (1974, p. 345). By Posner's reasoning, the coal industry should have welcomed the opportunity for industrial self-management.

The second inference is that significant divisions exist in the coal industry, divisions which may split the industry on particular issues. In fighting the NRA, presumably the "southern wing of the industry" and its allies had a common interest and sufficient individual incentives to behave as they did. Recalling Olson's arguments about collective action problems, the obvious reason for the coal industry acting according to the predictions of economic theory is either lack of common interest in stabilization measures or insufficient individual incentives. Johnson's characterization of the coal industry as "splintered" suggests that important economic and financial differences did exist among firms. These differences are the focus of this chapter.

Categorizing Coal Firms

The economic model of firm behavior points to profit functions as the determinant of regulatory impacts and firm preferences. Theoretically, each coal firm has a distinct profit function and, by extension, a unique view of any proposed regulation. Different profit functions logically imply the expression of different preferences. Though this sort of formulation makes sense from the standpoint of pure microeconomic analysis, it offers little help for making broad empirical statements about SMCRA's impact on firm preferences. As a practical matter the thousands of coal firms do not express thousands of individual preferences with regard to any regulation. Their judgments are not so refined. Rather, rational coal firms express a limited number of preferences, though they are guided by their "unique" profit functions. Logically, then, coal firms may be categorized according to profit functions. The most obvious dimension along which firms may be divided is size. Different sized firms possess different profit functions and hence, exhibit different preferences.

The National Coal Association (NCA) has adopted a particular method of classification that is highly sensitive to firm size. NCA divides coal firms into four categories: captive firms, interstate commercial firms, intrastate commercial firms and single unit producers. Although these categories also reflect the dispersion of a firm's operations and ownership, size is the primary distinguishing factor. At the time this scheme was developed, dispersion and ownership correlated almost perfectly with size.

Interstate commercials traditionally have been the giants of the coal industry. They mine in several states across the country, selling to foreign and large domestic customers. Most of these firms are well-recognized names in coal mining, and operate underground mines as well as surface mines. Captives are wholly owned subsidiaries of major noncoal companies that purchase the coal for their own internal use. These parent firms include primarily electric utilities and steel producers. All captives historically have been large relative to all but interstate firms since parent companies are major coal consumers. Intrastate firms are small-to-medium in size and supply some export as well as domestic markets. As indicated by the category name, they mine and usually market coal in one state. Finally, single unit producers tend to be very small and concentrated in the East, specifically Appalachia. By definition they too are intrastate firms, though this classification system intentionally categorizes them separately. They may sell coal to private or industrial customers, but more frequently contract with larger firms better equipped to market the coal. This type of mining firm in the past also has filled spot demand for coal in times of undersupply.

Although firm size plays an important role in distinguishing profit function, the NCA system or any method relying on size no longer suffices to differentiate firms according to regulatory impact. Recently, in the West, captives or even intrastate firms have begun to rival and, in some instances, have surpassed interstate commercials in size. This, of course, undermines the relationship between the NCA categories and size. Simply dividing coal firms into an arbitrary number of size categories obscures other important factors which may not correlate exactly with size, but rather more closely with geography. Two large firms may have quite different preferences with regard to SMCRA because their production functions and therefore their profit functions differ. The location of firm operations can account for these differences.

The Coal Industry by Region

An accurate description of American coal mining must differentiate the industry into three geographical regions: eastern, midwestern and western. Significant distinguishing characteristics among coal firms are coextensive with these regional divisions. The three regions differ in terms of topography, hydrology, rainfall, vegetation, wildlife and agronomy. In addition, the qualities of the coal, basically BTU's and sulfur content, vary from region to region as do the thickness and depth of coal deposits. Transportation facilities, marketing opportunities and landholding patterns also vary across the three regions. These factors all play a part in determining the nature of coal mining operations and land reclamation. Coal surface mining in any one of these regions constitutes a fundamentally different kind of enterprise as a result of these differences. Thus, the three geographical regions offer a convenient organizing principle for discussing the coal industry.

The East

The eastern coal mining region encompasses the states of Alabama, Kentucky, Maryland, Pennsylvania, Tennessee, Virginia and West Virginia. Beginning in 1975, the introduction of coal production in Georgia added that state to the list. Most of the coal in this region has been and remains mined by underground techniques. Surface techniques, though, account for a significant proportion of coal production. In 1979, of an estimated 420,388,000 tons of eastern coal, 181,854,000, or 43 percent, were extracted by some method of surface mining (calculated from Keystone Coal Industry Manual, 1980). This shows a modest increase over the 1972 figure of 37 percent (calculated from U.S. Minerals Yearbook, 1972). Much of the increase is due to new mining begun in the mid-1970s in Georgia and Alabama, the overwhelming percentage of which is surface mining.

Perhaps the most telling fact about the eastern coal industry is the tremendous number of small mines and firms, most of which are surface operations. Nineteen seventy-two is the last year in which the Bureau of Mines reported data on coal production and number of mines by mine size. In addition, the regional structure of the coal industry in 1972 is relevant since that is roughly the time SMCRA became a serious question on the national agenda. The BM broke mine size down to six tonnage classes: 0 – 10,000; 10,000 – 50,000; 50,000 – 100,000;

Table 4.1   1972 distribution of eastern coal mines by size and method
(thousand short tons)

Size	Surface mines	All mines	%	Surface as % of all mines	Surface as % of total surface mines	Class as % of total
Large	33	164	20	0.8	1	4
Medium	268	604	44	6.0	11	14
Small	2,136	3,546	60	50.0	88	82
Total	2,437	4,314	56		100	100

Source: *1972 U.S. Minerals Yearbook.*

100,000–500,000; 500,000 and over. For our purposes we can employ
the categories: "small"—up to 100,000; "medium"—between 100,000
and 500,000; "large"—500,000 and over.

As table 4.1 indicates, only 4 percent of all eastern mines were classified as large, while 82 percent were classified as small. Moreover, the
majority of all eastern mines were small or medium surface mines, 50
percent alone being small. Those small surface mines accounted for 36
percent of all surface production and 18 percent of total production. If
we add in medium-sized surface mines, the figures increase to 72 percent and 54 percent respectively.

In sum, the eastern coal industry includes numerous small and
medium-sized mines that produce significant proportions of the region's
coal. We also may infer that thousands of small and medium individual
coal firms operate in the East. This is as true in 1985 as in 1972. The
great number of small firms is largely a consequence of the ecological
features of the eastern region.

Topography is the single most important feature of the region from a
surface miner's perspective. The overwhelming majority of coal mining, underground and surface operations, takes place in relatively
mountainous parts of Appalachia. To be precise, Pennsylvania, Maryland and western Kentucky are not as mountainous as the rest of the
region. Nevertheless, on the whole, rugged terrain sets the East apart
from the other two coal mining regions. A generally accepted standard
of difficult surface mining terrain is a mountainside slope of 20 degrees
or more (ICF Study, 1975; Goldstein and Smith, 1975). These conditions obtain in over half of the eastern region according to a 1972 CEQ
report. In some critical high output areas such as southwestern Virginia

and eastern Kentucky, almost all surface mining is on slopes of 20 degrees or greater (Hearing before the Subcommittee on Energy and the Environment of the House Committee on Interior and Insular Affairs, 1977a, p. 12).

A second important characteristic of eastern coal mining is the size and depth of the deposits. In more technical terms, coal lies in "seams" or layers underground. In the East, frequently these seams lie one on top of the other, from the base of a mountain to a summit. Obviously, the thickness of a seam is a crucial variable in deciding whether it would be economical to mine that seam. In the East, coal seams are relatively thin, ranging from approximately 18 to 36 inches (House Hearings, 1977b, p. 20). An additional consideration is the depth of the coal, or the amount of "overburden" lying above it. A potential surface mine may be assessed in terms of the relationship between seam thickness and overburden, a relatively high ratio being more favorable. This ratio, for the eastern region, runs on the order of 1:18–20. By way of comparison, the ratio for other regions may run as high as 1:3 (Energy and Environment Analysis, Inc., 1977). Along with topography, these characteristics dictate certain methods of surface mining in the eastern region.

In general, three methods of surface mining are compatible with the eastern topography, that is slopes of at least 14 and frequently more than 20 degrees. These methods are contour mining, augering and mountaintop removal. Contour mining entails cutting a series of terraces into a mountainside and removing the coal from each terrace. The terrace cuts correspond to the coal seams as one moves up the mountain. The "spoil" or removed overburden is pushed over the mountainside into a valley or hollow below. When the mining operation is completed a "highwall" of up to one hundred feet remains against the face of the mountain. The highwall sits on a terrace or "bench" at a 90-degree angle. The actual procedure calls for the use of dynamite, a bulldozer and a small backhoe.

Augering is practiced when a coal deposit lies deep within a mountain. After cutting a terrace as in contour mining and removing the accessible coal, an operator bores into the mountainside to extract additional coal. As in contour mining, the spoil is pushed over the mountainside. Mountaintop removal can be practiced when the coal deposits lie on the upper one-third of a mountain. In this event, the miner literally removes the entire mountaintop, leaving a lower plateau. The advantage of this method is that nearly all the coal in the deposit is recoverable. On the other hand, this procedure requires significantly more men and machinery than either contour or auger mining.

In order to utilize any of these techniques, a mining firm must acquire the land rights to mine the coal. In the East, the pattern of ownership is highly diffuse. Numerous private individuals as well as coal companies own the land on which coal is mined. If the coal company cannot buy the land it needs, it must lease it on a long-term basis. In either case, this usually entails contractual arrangements with a number of individual owners for any sizable mine.

The combined impact of topography, seam-to-overburden ratios, mining methods and ownership patterns is to limit the size, expand the number and constrain the profitability of eastern mines. A single mountain with relatively thin seams attracts small independent surface miners. According to one industry source: "The existence of large reserves of high quality coals will tend to attract the investment which will make possible large mines and may result in large multiunit mines . . . mining firms that can only get smaller poorer quality coal deposits stay small or shrink" (Schmidt, 1979, p. 146). In the eastern region, the size of deposits is the operative factor in there being so many small coal firms.

Entry into the market is easy if all you need for a contour operation is dynamite and two or three moderate-sized earthmoving machines. Not surprisingly, a common practice in the East, at least in Appalachia, is for lumber or construction firms to operate single surface mines as a sideline. These surface mines produce 10,000 tons/year or less. Even larger mines producing 10,000 to 100,000 tons/year do not require bringing large amounts of capital and labor to bear on a surface mining project. Most large eastern operations, those producing over 500,000 tons/year or more, are underground or mountaintop removal mines run by big interstate commercial firms. Although Appalachia contains large enough deposits to attract these major firms, much of eastern coal is situated such that it attracts smaller surface operators. This led the President's Commission on Coal to describe the eastern coal industry as "a highly competitive industry characterized by low barriers to entry, many healthy small firms and small mines and a modest degree of economic concentration." These smaller mines are exclusively surface operations, in all likelihood passed up by major coal companies.

One factor favorable to eastern coal mining is the quality of the coal itself. Excluding the anthracite found only in parts of Pennsylvania, there are three grades of coal: bituminous; sub-bituminous; and lignite, or brown coal. These are listed in descending order of their BTU content. Most Appalachian coal is bituminous, and most of America's bituminous coal is located in the East. The other important quality characteristic of coal is its sulfur content. Since the establishment of the EPA, air pollution standards have placed a premium on low sulfur coal. In the East, sulfur content varies considerably, from 0.7 percent to 4.0 percent. On the whole, eastern coal is neither as clean as western coal nor as dirty as midwestern coal. The comparatively high quality in terms of BTUs tends to offset any problems with sulfur content in comparison with the other regions, though strict environmental standards have caused eastern utilities to use oil or natural gas. An ordinary BTU count for eastern coal would be 13,200/lb. of coal, a figure we can use for comparison with other regions.

A second important advantage of eastern coal mining is its transportation and marketing network. Because the coal industry developed in the East, the region has, in place, an extensive system of transportation geared to handling coal. Coal may be transported by railways, inland waterways or by sea. The eastern industry can employ all of these separately or in combination. River traffic and rail lines, for example, move coal north and south in Appalachia. For large industrial and utility markets of the Northeast as well as export, coal is moved by rail to Norfolk or Baltimore and up the Atlantic Coast or to Europe. In addition, long-idle coal docks in Philadelphia, northern New Jersey and Boston are now being refurbished. This network of rivers, ports and rail lines makes it relatively easy to move coal to any part of the region, Europe, or even the Midwest.

The element missing in this discussion, and in the minds of most eastern coal firms prior to the introduction of SMCRA, is the environmental impact of surface mining coal. A major reason for the survival of a large number of small surface mining firms in the East is the low capital and labor costs. These low costs are predicated on the limited expenses of eastern surface mining technology. However, the methods of surface mining, particularly contour and auger mining, have wreaked havoc on the environment of the eastern region. In describing these two methods, it was noted that the spoil was pushed off the bench and down the mountainside. This practice, though minimizing costs, creates two important problems. First, it leaves an exposed bench and highwall along the mountainsides. These areas are highly susceptible to erosion and landslides. Some estimates place the amount of exposed highwall in Appalachia at over 25,000 miles (House Hearings, 1977a, p. 53). This is the most obvious and ugly inheritance bequeathed by surface mining in the East.

An equally dangerous impact stems from the accumulation of spoil dumped in the valleys and hollows below the benches. This has led to the pollution and serious disruption of numerous small streams and rivers critical to the Appalachian watersheds. The pollution stems from sedimentation and the high acidity of the spoil from subsoil levels. In fact, even if surface miners retain the spoil on the benches to cover the highwalls, they must be careful to segregate topsoil and replace it on top of the subsoil. If they do not, rainfall will result in acid runoff into the valley streams below. Mountaintop removal obviously mitigates these problems of erosion from benches and highwalls. But it too can result in acid runoff without proper handling of the soil for reclamation. Poorly constructed mining roads on mountainsides also contribute to these problems of pollution and erosion. Clearly, without careful reclamation techniques and due attention to the hydrological impacts of surface mining, any hope of eventual ecological restoration is vain.

The Central Region

Coal mining in the central, or midwestern, region occurs in Arkansas, Illinois, Indiana, Iowa, Kansas, Missouri, Ohio and Oklahoma. Together, Illinois, Indiana and Ohio account for roughly 90 percent of the coal tonnage in this region, 129,250,000 of a total 142,470,000 tons in 1979 (calculated from state data in the Keystone Coal Industry Manual, 1980). This proportion has remained constant for at least the past fifteen years. In the central region, 96,300,000 tons, or 68 percent, of midwestern coal were derived from surface mines in 1979. This represents essentially no change from 1972 when the Bureau of Mines data indicated 63 percent, or 102,274,000 tons, was surface mined in the region.

Unlike the East, the midwestern coal industry is composed primarily of surface operations. As table 4.2 illustrates, a signficant percent of mines in all classes are surface operations. As in the East, small and medium mines tend to be surface mines. However, in the Midwest, over 60 percent of large operations are surface mines. More important, coal production is more heavily concentrated among larger firms in the Midwest. Large mines produced 82 percent of midwestern coal. Large surface mines alone accounted for 52 percent of all midwestern coal. This pattern holds true today. In the Midwest, major national companies carry out most of the mining both in terms of tonnage and number of mines. Again, the peculiar features of the region

Table 4.2   1972 distribution of midwestern coal mines by size and method (thousand short tons)

Size	Surface mines	All mines	%	Surface as % of all mines	Surface as % of total surface mines	Class as % of total
Large	54	89	61	12	14	20
Medium	72	89	81	16	19	20
Small	258	276	93	57	67	60
Total	384	454	85		100	100

Source: *1972 U.S. Minerals Yearbook.*

help to explain the size and nature of coal firms' operations.

Whereas the East is typified by fairly mountainous terrain in key coal fields, the central region has few slopes over 10 degrees. Coal fields in the Midwest lie on gently rolling land except for some hilly areas of Ohio and southern Indiana. Coal deposits extend good distances over relatively accessible terrain. Consequently, mining operations tend to cover greater areas.

Consistent with the greater extent of midwestern coal deposits, the other characteristics of the deposits are relatively favorable to large mines. In the Midwest, the size of coal seams ranges between four and ten feet in depth (President's Commission on Coal, 1980). The seam-to-overburden ratios tend to be much higher than in the East, on the order of 1:10 (Energy and Environment Analysis, Inc., 1977). In addition, the composition of the overburden makes digging the coal out much easier; topsoil is deeper and the ground is less rocky. Finally, the Midwest has its unique pattern of land ownership. Although lots are privately owned, they are large. This is due to both the topography and huge farms in the region. It is far easier than in the eastern region to acquire land or mining rights for extensive areas.

Predictably, the methods of surface mining differ from those employed in the East. Area mining and open pit mining predominate, though some parts of Indiana and Ohio require contour techniques. Area mining involves making a series of long trenchlike cuts in the land. The spoil from succeeding cuts is placed in the preceding trenches. This leaves a wave effect on the terrain since the spoil cannot be compressed to its original volume. Open pit mining, a less frequently used method, entails precisely what the name implies, excavating a single large pit in the coal field. This approach is warranted only with large and highly concentrated deposits sometimes occurring in two or more seams lying one above the other.

Extensive coal deposits, topography uninterrupted by mountains, comparatively thick seams and manageable overburden as well as fewer landholders all tend to make midwestern surface mining firms larger and more profitable than their eastern counterparts. The size of the mines themselves requires significantly more and heavier machinery. Power shovels, large trucks and bulldozers, all unnecessary in much of Appalachia, are essential in the Midwest. This situation acts as a barrier to entry for small operators. Deep mining start-up costs absolutely exclude small operators. Midwestern coal mining, surface or under-ground, is done by interstate commercial firms making relatively large initial investments and operating mines continually for periods of ten to twenty years. This contrasts markedly with the numerous small opera-tors in the East, many of whom fill spot market demand.

The more favorable character of midwestern coal deposits is partially offset by their lower quality. In particular, the sulfur content is exceed-ingly high, on the order of 5–8 percent. Moreover, BTU output, while not low, does not compare favorably with much eastern coal. Bitumi-nous deposits in the central region typically put out 12,000 BTUs/lb. There are also significant sub-bituminous deposits with still lower out-puts. In general, central region coal is dirtier and of slightly lower heat quality than eastern coal. Understandably, EPA standards on air quality under the New Source Performance Standards (NSPS) program hit mid-western coal especially hard. Midwestern legislators softened the impact somewhat when they succeeded in directing EPA to implement NSPS II which, rather than setting sulfur content limits at 1.2 lbs. SO$_2$/million BTUs, mandated percentage sulfur reductions in all coal no matter its original sulfur content or SO$_2$/million BTU output. Still, dirty coal remains a problem, accounting for lower Midwest output (see Ackerman and Hassler, 1981).

Like the East, the Midwest enjoys a relatively well-developed coal transportation system. Though its rail lines are not as numerous as in the East, the central region does have adequate rail facilities to service its utilities and metallurgical industries. The midwestern network of major rivers also provides an important transportation advantage, and the system of ports and shipping on the Great Lakes offers a low cost means of moving coal around the region.

Notwithstanding the political success in easing the impact of air pollution regulations, midwestern coal confronted strong environmental efforts to regulate its strip mining activities. The environmental impact of central region surface mining, while severe, is not as striking as in the East, at least to the casual observer. The land is not scarred by high-walls. However, problems of erosion and water pollution still occur. These impacts are exacerbated in the Midwest by the fact that much of the disturbed land is taken from highly productive farmland. Serious problems result from open pit mines in which the spoil is dispersed above and around the pit. Even the area mines, though, have serious impacts. Still, unless the topsoil stripped off is segregated from the rest of the spoil and replaced on top, problems of acid runoff and siltation can prevent the land from being returned to farming. Construction of mining roads at large mines contributes to soil management problems as well.

The Western Region

The western region encompasses all coal mining west of the 100th meridian. The coal states in this region are Alaska, Arizona, Colorado, Montana, New Mexico, North Dakota, Texas, Utah, Washington and Wyoming. By far the largest coal reserves lie in the Northern Great Plains states of Montana, Colorado, Wyoming and North Dakota. To-gether these four states produce about 70 percent of the region's tonnage, 140,470,000 of a total 205,710,000 tons. In the West, mines extract coal almost exclusively by surface techniques: 184,170,000 tons, or 90 percent of western coal, is surface mined. Only Colorado and Utah have any deep mines (calculations from Keystone Coal Industry Manual, 1980). Surface mining's share of western tonnage remained fairly stable through the mid-1970s. Even in 1972, surface mining accounted for 83 percent.

Turning to the question of mine and firm size, the West presents a picture the reverse of the East. In the western region, extremely large mines are the rule, and almost all large mines are surface mines (see table 4.3). Surface mines appear even more predominant when we consider that underground mining is concentrated in Utah.

Although there were a number of small and medium mines in 1972, 94 percent of western tonnage was produced by large mines and 77 percent by large surface mines. Indeed, by 1980 even the numbers of mines had tilted in favor of larger operations. Because the size of the capital investment for a coal mine must match the scale of operations,

Table 4.3 1972 distribution of western coal mines by size and method (thousand short tons)

Size	Surface mines	All mines	%	Surface as % of all mines	Surface as % of total surface mines	Class as % of total
Large	24	27	89	22	47	27
Medium	10	34	29	9	20	34
Small	17	50	34	15	33	45
Total	51	111	45		100	

Source: *1972 U.S. Minerals Yearbook.*

western firms tend to be large and have ready access to financing. As in the East and the Midwest regional factors dictate the size and nature of coal firms.

Topography presents almost no problems in the West. The terrain in which the coal deposits lie is even gentler than that in the central region. Moreover, very little land on which coal is found has agricultural uses. Of this agricultural land, most is rangeland used for cattle grazing, not prime farmland. The main attraction of western coal deposits, however, is their size and depth. Western coal seams can extend over several square miles and range in depth from 20 to 100 feet or more (President's Commission on Coal, 1980). The layers of overburden in the West are as easily managed as those in the Midwest. Seam-to-overburden ratios vary from 1:8 to 1:3, the higher ratio obtaining in the rich deposits of the Northern Great Plains (Energy and Environment Analysis, Inc., 1977).

As if these advantages were not sufficient, the sulfur content of western coal falls uniformly below that of either the eastern or central regions. After the enactment of the Clean Air Act and EPA's promulgation of NSPS, low sulfur content proved to be a significant economic advantage. While it is true that the BTU output of western coal lags behind the East and the Midwest, 8,500–10,600 BTUs/lb. for subbituminous western coal and 7,200 BTUs/lb. for the lignite deposits, the low sulfur content and size of the deposits makes western coal an attractive energy source.

Area mining and open pit operations, primarily the former, predominate in the West. Western mines, though, are immense compared to similar operations in the central region. This difference dictates a slightly different approach and significantly higher costs. Western surface mines require gigantic earthmoving equipment capable of handling tens of tons of overburden or coal at once. Large mine size follows logically from the defining features of the environment and coal deposits.

Ownership patterns in the western region have presented some difficult, though not insurmountable, problems in this otherwise favorable setting. In the West, a coal firm ordinarily must deal with two, and sometimes three, different kinds of landholders in acquiring access to coal—private owners (usually ranchers), Indian tribes and the federal government. Though Indian ownership falls under the Omnibus Tribal Leasing Act of 1937, we may treat the tribes as private owners for our purposes. Federal ownership, however, complicates acquisition significantly. To set this issue in proper perspective, of the 214.62 billion tons of proven reserves in the western region, 128.77 billion tons, or 60 percent, lies on federal lands (President's Commission on Coal, 1980). Federal ownership became a stumbling block for coal development in the region when environmental concerns became firmly established on the national agenda.

Western mineral deposits on public land were allocated under the Mineral Leasing Act of 1920. Administered by the Department of Interior, this law served as a vehicle for developing various mineral resources in the West. Until 1945, though, very little coal was leased under this act. Coal leasing activity accelerated in the 1960s when farsighted individuals recognized that by the end of the decade the huge machinery necessary for surface mining western coal would be available. Sixty-seven percent of all western leases were issued between 1960 and 1971, and the percentage in terms of tons of coal is even higher (Senate Hearings, 1974, p. 224). In 1971, BLM issued a study revealing that "the acreage of coal under lease in the public domain was skyrocketing while production on Federal leases had declined . . . since 1945." In 1945, 80,000 acres of public coal land was under lease and 10,000,000 tons were produced while in 1970 778,000 acres were under lease but only 7,200,000 tons were produced (Senate Hearings, 1974, p. 95).

Following the publication of this information, DOI placed a moratorium on all western coal leasing while it investigated the charges of speculation. Under the 1920 leasing law, coal land had to be leased under competitive bidding and the successful lessee had to "continuously develop" the coal deposit. One qualification to this provision was that the Secretary of Interior could issue, at his discretion, "preferential" leases to individuals or firms to help establish unified ownership of large deposits. Because it might take some time to negotiate the rights to an economic sized mine, coal firms could pay one year's advanced rent to DOI in lieu of continuous development. These rents averaged only $200 dollars per annum on some leases.

This system invited speculation by coal brokers, a possibility that DOI, in 1971, decided to foreclose. Although speculation was the main concern of DOI and various consumer lobby groups, environmentalists more concerned with the noneconomic impacts of western surface mining strongly backed the moratorium. When it became apparent that DOI would continue leasing under an amended procedure, environmental groups, also pushing for SMCRA, sought to influence that program. The ultimate outcome of their efforts was the Federal Coal Leasing Amendments Act of 1976 (P.L. 94-377).

The period 1971 to 1976 saw the institution of several DOI actions to eradicate speculation, all of which bore the stamp of environmental concerns about western coal mining: (1) December, 1970—BLM imposed the moratorium on coal mining; (2) December 17, 1973—Interior Secretary Morton promulgated an interim leasing program based on an Energy Minerals Allocation Recommendation System (EMARS) under which the federal government determined the amount of leasing through its own demand and supply analysis; (3) June 6, 1973—DOI Memo 73-231 directed BLM to reject all pending lease or mine applications, to consider them under EMARS and to require an EIS; (4) August, 1973—DOI developed the Coal Programatic Environmental Statement (CPES) to be completed by all firms requesting leases. Even though a concern about speculation under the 1920 law prompted the initial moratorium, environmentalists worried about rapid and uncontrolled development of western coal resources clearly left their imprint on these subsequent DOI actions.

Because of the need for large mines in order to make investments worthwhile, western operators inevitably encountered problems of requiring one or two more leases to make a mine go. Frequently these leases, when owned by the public, were not forthcoming in the early 1970s. After 1973, firms had to adjust to the requirements of EMARS and CPES, further delaying start-ups. The lead time, even before all this federal involvement, ran to five years at a minimum. Therefore, the lands leased in the mid-sixties could not have become productive until the early 1970s, precisely the time of the policy confusion. In this milieu, western coal firms had to develop a capacity to deal with the federal government and environmental concerns. This factor, in combination with other features of coal in the West, restricted the region to only very large firms with considerable financial and managerial resources.

Transportation poses another problem for western coal mining. Major deposits of western coal lie in the Northern Great Plains and areas of New Mexico and Arizona, far from most industrial and public utility consumers. More important, the transportation system in the West is relatively undeveloped. In part this is due to the fact that the region is not blessed with the navigable rivers and natural ports found in the other two regions. The transportation of western coal depends almost exclusively on rail traffic. While the western region has a number of major freight lines, opening a coal mine ordinarily involves building a rail spur from the mine to a trunk line. This cost is usually assumed by the mining firm and passed on to consumers as much as possible. These construction costs plus the great distances over which the coal must be moved add significant increments to the cost of western coal.

In addition to the "natural" factors which increase the cost of marketing western coal, firms must contend with the ICC, which has sanctioned exorbitant rail rates. A case involving COLOWYO Coal Company and the Central Texas Power Company (CTP) illustrates the ways in which the ICC and railroads can impinge on western coal mining. COLOWYO contracted with CTP to deliver coal for a thirty-year period from a new mine in Axial, Wyoming. In preparation, COLOWYO built a rail spur to transport the coal to a trunk line and CTP purchased 1,500 hopper cars. It cost COLOWYO $7.00/ton to get the coal out of the ground and loaded on trains. However, the only rail line operating between Axial and CTP plants charged a tariff of $20.00/ton. Although the tariff was appealed, the ICC ruled that it could stand since an alternative means of transportation existed. Incredibly, the ICC held that COLOWYO could honor its contract by moving the coal by rail to the Pacific coast, shipping it by sea to Galveston, and then sending it by rail again to central Texas. Confronted with this choice, COLOWYO and CTP voided their contract. While this example is extreme, it illustrates the potential impact of transportation problems on western coal mining.

In an attempt to deal with the rail monopoly, coal firms in the western region have explored other transportation options. They have studied the possibilities of constructing coal slurry pipelines or power stations at the mouth of mines. Slurry pipelines are pipelines that pump coal along with water. One coal pipeline already operates in New Mexico at Western Energy Corporation's Black Mesa Mine. The transportation savings are significant. The greatest savings, however, would accrue from moving coal across state lines: the most ambitious project is to construct a slurry pipeline from the coal fields of Gillette, Wyoming, to Baton Rouge, Louisiana. This project, proposed by a consortium, Energy Transportation Systems Inc., necessarily involved the federal government in acquiring the right of eminent domain to build the pipeline.

Railroads controlling easements vital to the project, though, have vigorously opposed the pipeline in Congress, as has the ICC. The opposition has succeeded thus far. It should be noted that environmentalists opposed the project also on the grounds that it would deplete the sensitive water tables of the West and promote large-scale surface mining in the region. This situation occasioned a curious coalition of environmentalists and business interests. The other option, building power stations at the mine site, has not yet proven economical.

Surface mining in the West presents its own unique environmental problems. The most serious consequence of coal mining in this region is the effect on hydrological resources. Throughout the region, water is scarce. The area and depth of western mines can disrupt or pollute underground aquifers vital to regional ecologies. In addition, revegetation is difficult since the average rainfall in the West is quite low. Another concern of environmental groups and western citizens is that surface mining may do irreparable damage to the scenic beauty of the region. Although some environmentalists' fears of a second Appalachia, scarred almost beyond recognition, are unfounded, the sheer magnitude of western surface mining operations presents extremely difficult problems for effective reclamation.

Summary and Trends

In a sense, three separate coal mining industries exist in the United States. The East, the Midwest and the West each has its own particular features which proscribe the mix of surface and underground mining as well as the scale and the methods of mining. Superimposed on this regional configuration, there is a fourth branch of the industry encompassing those firms that mine coal in more than one region. Such firms, operating for the most part in the West and one other region, must employ policies and practices that accommodate all the regions in which they mine coal. None of these four "industries" has remained static over the past ten to fifteen years, the period in which SMCRA was enacted and implemented.

Table 4.4   Acquisitions of coal companies by multinational conglomerates over a ten-year period

Oil firm	Acquired firm	Date
Gulf Oil	Pittsburgh & Midway	1963
Continental Oil	Consolidation Coal	1966
Occidental Petroleum	Island Creek Coal	1968
	Maust Properties	1969
SOHIO	Old Ben Coal	1968
	Enos Coal	1968
Eastern Gas & Fuel	Joanne Coal	1969
	Ranger Fuel	1970
	Sterling Smokeless	1970

Source: Ridgeway, 1973.

One change with extremely far-reaching implications has been the entrance into the coal industry of multinational conglomerates, especially petroleum companies. In the late 1950s, oil firms attempted to acquire coal reserves, primarily in the western region. However, owing to the amicable relationship between coal firms and leasing authorities at DOI, this proved difficult. As a result, many oil firms interested in long-term plans for coal gasification, shale oil or simply diversification in the energy field sought to acquire coal firms. The first major acquisition occurred in 1966 when CONOCO, then Continental Oil, acquired Consolidation Coal, the nation's second largest producer. "The Continental-Consolidation acquisition was financed by production payment arrangements, under which a loan to finance the acquisition was secured by the proceeds of the acquired firm's future production" (Schmidt, 1979, p. 146). This set a pattern for future acquisition of large coal companies. Table 4.4 lists important acquisitions over a ten-year period. The involvement of oil has continued into the 1970s as Exxon acquired Monterrey Coal and Carter Coal as a base for its coal operations.

By 1980 all major oil firms were into the coal industry. Most recently (March, 1981), SOHIO tendered an offer for AMAX Coal, an Indiana-based interstate firm traditionally among the top ten to fifteen producers. Between 1962 and 1970, the market share of independent coal companies dropped from 31.5 percent to 9.7 percent. In that same period, the share of oil and natural gas firms jumped from 2.1 percent to 23 percent, with other large noncoal firms increasing their share from 2 percent to 15 percent (Schmidt, 1979).

It was no coincidence that the move of big oil into big coal coincided

Table 4.5 Western shift in growth of U.S. coal production from 1975 to 1979

1975			1979		
Company	State	Tonnage (millions)	Company	State	Tonnage (millions)
Decker Coal	Montana	9.2	AMAX	Wyoming	15.0
Peabody Coal	Illinois	6.5	Decker (1)	Montana	13.0
Western Energy	Montana	6.4	Western Energy	Montana	11.7
Utah International	New Mexico	6.1	Texas Utilities	Texas	10.5
Peabody Coal	Kentucky	4.6	Texas Utilities	Texas	10.0
Peabody Coal	Kentucky	4.3	Peabody (7)	Arizona	8.0
Peabody Coal	Illinois	4.1	Thunder Basin	Wyoming	6.2
Westmoreland Resources	Montana	4.0	Utah International	New Mexico	6.0
Pittston Company	Virginia	3.8	Bridger Coal	Wyoming	5.7
			Peabody (2)	Illinois	5.4

with the expansion of western surface mining. It was a happy marriage between oil capital and coal reserves. Oil companies could pursue their plans for gasification or diversification, both of which required access to large coal mines in the East or Midwest, but more important, the untapped reserves in the West. The coal firms would have had great difficulty affording the move into western surface mining on their own, though they had the expertise. The westward shift of coal, incidentally, made huge older firms less vulnerable to the vicissitudes of UMW-management problems associated with other regions. The nonunion character of western labor also attracted the oil firms.

To get a crude idea of the westward shift in the industry, we can note in table 4.5 the difference in the location and output of the top ten coal mines in the United States at the beginning and end of the five-year growth period from 1975 to 1979. For 1979, numbers in parentheses next to mining companies indicate their rank in 1975. Only one non-western mine appears on the 1979 list. Indeed, that Illinois mine, Peabody's River King, is the only nonwestern mine in the top twenty producers of 1979. This shift is striking.

Table 4.6 presents this shift toward western coal mining in annual production figures and regional percentages. In addition, there was a general shift across all three regions to surface mining in the same period as indicated in table 4.7. Comparing tables 4.6 and 4.7, it is easy

Table 4.6   Coal production by region, 1965–1977 (thousand short tons)

Region	1965	1966	1967	1968	1969
East	73%	72%	71%	71%	70%
	371,416	382,746	393,422	388,088	394,928
Midwest	24%	24%	25%	25%	25%
	120,364	129,895	137,022	134,091	139,857
West	3%	4%	4%	4%	5%
	20,308	21,240	22,182	23,066	25,720
Region	**1970**	**1971**	**1972**	**1973**	**1974**
East	69%	68%	65%	63%	63%
	417,846	373,582	387,249	374,797	377,719
Midwest	25%	25%	26%	25%	24%
	149,941	136,303	153,483	149,467	142,524
West	6%	7%	8%	12%	13%
	35,145	42,307	54,654	67,474	83,163
Region		**1975**		**1976**	**1977**
East		61%		60%	57%
		396,487		406,162	389,850
Midwest		23%		22%	21%
		151,116		147,892	146,365
West		16%		18%	22%
		100,835		124,631	152,360

Source: *Bureau of Mines Energy Information Administration Survey*, 1977.

to see that the westward shift coincided precisely with the increase in surface mining. Western expansion accounts for most but not all of the switch to surface techniques. Both trends fit the increasing concentration in coal mining. By 1975, 27 percent of the surface mines in the United States extracted 87 percent of all surface-mined coal (Shover, 1980).

A final important trend in the coal industry is the shift in consumption. Between 1972 and 1976 the percentage of coal for generation of steam electricity by public utilities grew from 24 percent to 70 percent of total consumption (U.S. Minerals Yearbook, 1976). In that same period, railroad consumption dropped from 10 percent to zero and retail sales dropped from 14 percent to less than 1 percent (ibid.). These shifts are consistent with the trend toward concentration. Large companies are signing 30–40-year contracts with electric power utili-

Table 4.7   Coal production by method, 1965–1977 (millions of tons)

Year	Underground	Surface	% Surface
1965	333	179	35
1966	339	195	36
1967	349	204	37
1968	344	201	37
1969	347	213	38
1970	339	264	44
1971	276	276	50
1972	304	291	49
1973	299	293	49
1974	277	326	54
1975	293	355	55
1976	295	384	57
1977	272	417	61

Source: *EIA Annual Report to Congress,* 1977.

ties. This trend, combined with the increasing control of reserves and production by large companies, has put even more pressure on the smallest operators. It also cushions the impact of SMCRA on the larger firms, which confront a problem of how to pass costs on to public utilities, not of how to remain competitive.

Conclusion

Based on the description of the coal industry, we can differentiate firms along two dimensions, size and region. For SMCRA or any other environmental regulation, regional differences and size differences matter. They determine how a particular firm will go about complying with uniform regulatory standards. These two factors shape the firm's profit function, its basis of preference formation. The profit functions of similarly sized and located firms are similar.

As a rule, region determines mine size and, therefore, firm size. Moving from east to west, size tends to increase and surface mining increases as a proportion, especially among large firms. Consequently, we should observe different profit functions in each region. Roughly, this is the case. Table 4.8 represents the capital investment (fixed costs) and operating expenses (variable costs) for typical surface mines in each region. It also sets forth typical yearly tonnages (quantity). Clearly there is a close relationship among mine size, region and profit function. To

Table 4.8   Capital investment and operating expenses for typical surface mines in each region (in yearly tonnages)

Costs	East (100,000 tons)	Midwest (500,000 tons)	West (5,000,000 tons)
Initial capital investment	$ 892,230	$17,501,000	$30,221,500
Deferred capital investment	2,667,330	10,087,800	22,425,500
Total capital investment	3,559,530	27,588,800	52,647,000
Operating cost per year	1,515,598	4,492,753	11,438,944

Source: President's Commission on Coal, 1979.

the extent that individual mines reflect firms' operations, we find a close relationship also among firm size, region and profit function. This relationship, however, can be deceiving.

If we hold firm size constant, we would not find similar profit functions among firms in different regions: this, after all, is the main point of our industry description. A major reason for this is the percentage of surface mining which varies across regions within a single class of coal firms. More direct regional factors such as topography or environmental factors also dictate different surface mining technologies and, hence, different profit functions across regions for the same size firms. Also, size alone will distort a comparison of interregional firms with large firms operating in a different region. On the other hand, holding region constant, we will note considerable variation among firm profit functions which can be best explained as a function of firm size. Coal firms ought to be characterized, therefore, in terms of a size dimension (small, medium, and large) as well as a regional dimension, though the latter must incorporate interregional firms.

Any coal firm may be categorized in this way. For interregional firms though a question arises: what does this category mean in terms of a profit function or regulatory impact? Answering this question requires knowledge of the regional mix of the firms' mining operations. In other words, where do they currently mine most of their coal, and where are they likely to mine it in the near future? This information should provide enough "insight" into the firm's profit function to determine its preferences with regard to SMCRA.

5   SMCRA and Firm Preferences

The analytical framework developed in chapter 2 indicates that an assessment of SMCRA's effect on business behavior should begin with firm preferences. A firm's preferences direct its behavior in regulatory politics. Accordingly, based on the preceeding discussions of SMCRA and the coal industry, this chapter derives a set of preferences for coal firms. This approach is intended to parallel the rational decision processes a coal firm would employ: evaluate SMCRA in terms of self-interest; form preferences based on that evaluation; determine behavioral options; participate in regulatory politics subject to political-economic constraints. In this chapter, we focus on the first two processes.

Given the assumption of rationality, preferences will be revealed through behavior (Riker and Ordeshook, 1973). Much of the evidence, such as congressional testimony and information from interviews, therefore, will demonstrate interfirm differences in behavior as well as preference. Nevertheless, in a conceptual sense it is important to treat preference formation as distinct from behavior even though for practical purposes there will be considerable overlap in the analysis of each. The analysis of preferences will lay the groundwork for a discussion of firm behavior and influence.

Preference Functions

The new regulatory regime and its particular subgovernmental elements such as SMCRA should evoke a variety of preferences from firms subject to regulation. This variation owes primarily to the uniform application of rules and requirements to firms in fundamentally different economic and financial positions; in other words, to firms with different profit functions. The description of the coal industry leads to specific expectations about such interfirm differences which, when appraised in light of SMCRA and its regulations, suggest specific firm preferences. The interfirm differences in profit function, we have seen, are summarized by size and region variables. Generally, distinct preferences should characterize coal firms in each of the relevant categories organized along the size and region dimensions. However, a question remains. How will distinct preferences manifest themselves?

At the most rudimentary level, firms may "favor" or "oppose" SMCRA. Each category of coal firm, therefore, could fall into a pro-SMCRA or anti-SMCRA camp. For some firms absolutely antagonistic to the new regulatory regime and to SMCRA in particular, the classification "opposed" may aptly represent their preference. Yet it would be simplistic to infer that all other firms "favor" SMCRA.

Any legal instrument as detailed and complex as SMCRA is bound to draw a range of responses from affected firms depending on which provisions and regulations impinge on them and with what severity. Also, since amendment, bargaining and compromise are endemic to the legislative process, room exists for a range of preferences beyond simple opposition or support for a bill. In addition, firm preferences may shift as the regulatory program moves from the legislative to the implementation phase. The problem, then, is how to actually distinguish among firm preferences.

To resolve this problem, we begin by noting that a discussion of preferences for rational actors presupposes the existence of preference functions, or rank-ordered sets of social outcomes. Since rationality implies purposive behavior, these outcomes serve as the objects of firm behavior. They must be specific and in accordance with the particular context of regulatory policy. For example, the totally antagonistic firms alluded to above would exhibit a dichotomous preference function: SMCRA is defeated or enacted.

In contemplating a regulatory bill as complex as SMCRA, most coal firms can seriously entertain a number of outcomes. To put this in more concrete terms, if there are three amended versions of SMCRA under consideration, each would constitute a separate outcome and contain a different combination of policy changes. For example: (1) complete defeat of SMCRA; (2) eliminate bans on prime farmland mining; (3) eliminate public hearings in the permitting process; (4) eliminate alluvial valley bans; (5) represent enactment of the original bill. Outcomes 2–4 represent the amended versions.

Now consider two large coal firms, one mining exclusively in the Midwest and the other exclusively in the West. The former should exhibit the preference function (1, 2, 3, 4, 5), while the latter should prefer the ordering (1, 4, 3, 2, 5). These two firms share an interest in eliminating public hearings and even accord that interest the same priority. However, SMCRA's particular impacts yield different preference functions. These kinds of differences are relatively easy to understand.

More complex considerations, though, may yield more complex preference functions. Legislative bargaining is likely to yield alternative versions of SMCRA with combinations of amendments unlike the simple case cited above. Three compromise versions of SMCRA might be (2) elimination of prime farmland bans and public hearings; (3) elimination of alluvial valley bans, public hearings and reduced reclamation fees; (4) elimination of alluvial valley bans and prime farmland bans.

A firm's operations, moreover, may be interregional. In the case of a Midwest-based interregional with substantial western surface operations, its preference function might look like (1, 4, 2, 3, 5). For this firm its midwestern operations are of such importance that it places the highest priority on eliminating the prime farmlands ban. Even though 3 offers more advantages than 2, eliminating the prime farmlands ban at any cost is the guiding principle for this firm.

The appropriate concept for understanding these kinds of preference functions is a lexicographic function. A lexicographic preference function is one in which an individual actor, in this case a coal firm, sets so much store by a particular social outcome that it will not consider compromise on other aspects unless satisfied on the first. The Midwest-based interregional discussed above prefers one version of SMCRA to another *if and only if* the first eliminates the ban on prime farmland mining. Other kinds of relief such as the elimination of alluvial valley bans or public hearings become relevant *if and only if* the prime farmland criterion is met.

Thus, any firm may desire many amendments to SMCRA, but one or some may be prerequisites to further considerations. As we shall see, lexicographic preference functions reflecting definite priorities characterize coal firms in their assessment of SMCRA. This is to be expected from a regulatory policy with regionally specific as well as general impacts.

Any variations in preference functions should be more pronounced as enactment of SMCRA becomes more certain and as the regulation moves into the implementation phase. In the early legislative phase, practically all firms should prefer no environmental regulation. However, once enactment seems assured, and especially once the implementation process begins, those firms with a dichotomous preference function will pursue their first preference in another arena, probably the courts, which historically have offered a second chance to losers in the legislative process. The other firms will abandon their first choice and participate in the development of implementing regulations with an eye toward easing the burdens of compliance. For all intents and purposes, they ignore the possibility of overturning SMCRA. At these later junctures, simple differences between antagonistic firms and other sharpen as do lexicographic differences among firms bearing different regulatory impacts.

In characterizing the preference functions of the various firm categories, the analysis, of necessity, will be less formal than perhaps implied by the preceding discussion. That discussion was intended to clarify an approach to understanding preference formation among rational coal firms. From the personal interviews and the public record, it is possible to reconstruct a rank ordering of major concerns for different firm categories. It is not possible to model, in a formal sense, firm preferences for a bill as complex as SMCRA. The number of alternative versions would be astronomical.

However, insofar as coal firms have limited capacities to assimilate and process information, an approximation of firm preference functions will suffice. As practical political participants, coal firms cannot thoroughly consider the hundreds of specific provisions of SMCRA. Rather, each coal firm will focus on a small number of high impact provisions from which it will form its preference function. Following the organizational format of the coal industry description in the preceding chapter, this discussion too will be structured by region.

The Eastern Firms

Small and medium-sized coal operators comprise the two relevant categories of eastern firms. With few exceptions, these firms exhibited fierce opposition to SMCRA in Congress. That opposition, moreover, carried into the implementation phase. Alterations in the substance of the program from the 1972 Hechler bill to abolish surface mining through the final version, H.R.2 passed by the 95th Congress, did nothing to change the basic antipathy of eastern surface mine operators.

Their preferences derived from the interaction of SMCRA's provisions and regulations with regional factors and firm size, the result being staunch and implacable resistance. In short, eastern surface mine firms exhibited the dichotomous preference function. SMCRA, in any form foreseeable under the political-economic realities of the new regulatory regime, would be unacceptable to eastern operators. First, environmentalists viewed eastern surface mining firms as major culprits in the depredation of natural resources in the East. Their view was supported by the tens of thousands of miles of exposed highwalls in Appalachia and the innumerable polluted streams throughout the East. Second, the size of eastern firms tended to make SMCRA's impact particularly severe. Since the overwhelming number of eastern firms and mines are small or medium-sized surface operations, high compliance costs for SMCRA fall more heavily on them.

Environmentalists and their congressional allies clearly tailored SMCRA to address environmental problems associated with eastern surface mining. More to the point, SMCRA took dead aim at the eastern operators believed, by environmentalists, to be the worst offenders. Representatives of the Citizens Coal Project expressed an entirely unsympathetic view of "recalcitrant eastern surface miners" (House Hearings, 1973c). Citizens Coal Project is a study and lobbying group interested in social impacts of coal mining. Sponsored by the Environmental Policy Institute (EPI), the Project had substantial input in the drafting of SMCRA.

This concentration on the eastern firms continued into the implementation phase. According to Dick Hall, director of enforcement at OSM and a co-founder of the Natural Resources Defense Fund, a significant departure in SMCRA is "the authority granted to those at the bottom of the enforcement pyramid." In Hall's view, the great advantages of this institutional change were that it prevented firms from "getting too close" to central regulatory authorities and mine inspectors could shut down, on the spot, small irresponsible surface operators in the East. Thus, potential environmental problems could be averted (personal interview, July 18, 1980).

This focus served as an orientation for those drafting SMCRA, predisposing them to take a hard line on those portions of the law that would regulate eastern contour mining firms. Throughout the legislative phase, pro-SMCRA forces showed no inclination to seek a political accommodation with eastern firms. In the implementation phase, two of the original eight performance standards comprising the interim program dealt specifically with eastern surface mining, while none of the remaining six singled out a particular region. Thus, the approach of environmentalists and their congressional allies insured opposition from eastern coal firms.

The provisions and regulations bearing most directly on eastern coal mining deal with steep slope mining. In particular, SMCRA requires that coal surface mining firms must return the land to its approximate original contour. Furthermore, it prohibits coal surface mining on slopes of greater than 20 degrees unless the firm receives a special variance from OSM. Such a variance is extremely unlikely for the majority of eastern firms that can afford only contour techniques rather than mountaintop removal. Given the mountainous topography throughout much of the East, especially in central and southern Appalachia, these provisions obviously focus on eastern surface mining. Neither provision was changed significantly from one version to the next of SMCRA. And, once the companion regulations detailing approved methods of compliance were written, steep slope provisions became even more onerous for eastern surface miners, primarily because they had limited resources.

The question of limited resources raises the second point about SMCRA's impact in the East: most eastern firms are small and medium-sized. In 1973, a Council on Environmental Quality report to the Senate, *Coal Surface Mining and Reclamation: An Environmental and Economic Assessment of Alternatives*, noted that "Much of Central Appalachian coal is from small operators who cannot afford capital and financing for compliance or switching to underground techniques" (CEQ Report Pursuant to S. Res. 45, 93-8, 1973b, p. 68). The economic health and, in many cases, survival of eastern surface mining firms were predicated on the use of contour techniques and minimal reclamation in very mountainous terrain. As the legislative debate on SMCRA wore on and it became obvious that there would be no meaningful compromise on the issues crucial to small and medium-sized firms, especially those situated in the East, their preference functions became locked in to the dichotomous form.

The aspects of SMCRA relevant to considerations of firm size are the incremental costs of reclamation and so-called front-end costs imposed by the law on all coal operators. Front-end costs include performance bonds, reclamation fees assessed per ton of coal production and expenditures for generating technical environmental data for an OSM surface mining permit. All of these front-end costs posed a formidable obstacle to coal mining for small operators. SMCRA, for example, mandates a minimum performance bond of $10,000 per mine compared to bonds as low as $25 per acre that were not uncommon in the East. For a typical small-to-medium-sized surface mine of forty acres, the earlier bond would require only $1,000 cash. Moreover, SMCRA provided that only 60 percent of the bond could be released after five years compared to the earlier practice of 100 percent returned in three years. As for reclamation fees, SMCRA originally was to impose a levy of 35¢ per ton of coal, this money to finance a fund for the reclamation of abandoned or "orphaned" mine land. Even the eventual levy of 15¢ per ton in H.R.2 amounts to an additional $7,500 per year for a mine extracting only 50,000 tons per year.

These expenditures present a significant cost for smaller mines facing uncertain markets and tight financing in the 1970s. An analysis of the coal industry concluded: "the uncertain market has put the squeeze on many small coal companies. And there is a winnowing process under way in the Eastern coal fields . . . the growing cost and complexity of Federal regulation may have made a comeback too expensive for many" (*Business Week*, September 24, 1979, p. 104). Corroborating this view, a DOE survey estimated that about one thousand small coal firms went out of business between 1965 and 1980. Significantly, over half that number left after 1977, the year of SMCRA's enactment. While the changing structure of the coal industry and coal markets played a role in this winnowing process, it is clear that SMCRA did too.

A number of estimates for permitting costs existed prior to the enactment of SMCRA, and all indicated a level of expenditure that would impose considerable financial stress on smaller firms. Experience with SMCRA after 1977 substantiated these preliminary indications. In hearings on the Senate version of SMCRA, S.7, an environmental engineering company reported to the Committee on Energy and Natural Resources that accumulating requisite data from core drillings and hydrological studies alone could result in permit costs of $36,000 to $330,000 for a 40-acre surface mine in the East (Senate Hearings, 1977c, p. 86). The House Subcommittee on Energy and the Environment, meanwhile, received an estimate that engineering services for that same geological and hydrological data would range from $50,000 to $100,000 (House Hearings, 1977a, v. 2, p. 166).

In both cases, the actual figure varied with the particular geological characteristics of the mine in question. Nevertheless, these costs represented significant expenditures for a smaller mine or firm. In its own cost study after enactment, Consolidation Coal, a huge interregional, estimated the cost of permitting proceedings at a mine in Appalachia at approximately 2¢/ton (Consol Study, 1979). However, a mine producing only 50,000 tons per year in the same region would face the same costs. More important, a firm like Consol can expense these costs through in-house engineering and scientific services, whereas the smaller firms must come up with cash to contract for the services.

Essentially the same situation obtains with respect to the incremental costs of reclamation. Incremental cost is the cost of any action to comply with SMCRA that would not have been taken in the absence of SMCRA. It is important to recognize that for most small and medium-sized eastern firms, SMCRA's reclamation costs would be a completely new expenditure. The 1973 CEQ study cited above presented the reclamation costs for a "typical" contour mine in the East, as shown in table 5.1.

It is not necessary to develop a technical discussion of these various reclamation techniques. Suffice it to say that SMCRA's provisions and companion regulations on return-to-approximate-original-contour required, in almost every case, contour backfilling. Thus, eastern operators faced an incremental cost of nearly $1.00 per ton. For the overwhelming majority of these firms, that meant a cost of $25,000 to $200,000 per mine per year. Any reclamation technique other than contour backfilling required filing for a special variance with OSM. Given the past mining practices and political resistance of small and medium-sized eastern firms to SMCRA, approval by OSM was extremely unlikely.

Reflecting on the content of the regulatory program, it is not difficult to see that SMCRA was geared toward concerns over the mining practices of eastern firms, especially the overwhelming number of small ones. If the permitting and bonding requirements did not put the worst offenders out of business, the reclamation fees might. If smaller firms could survive through financing these costs, the specific design and engineering standards written into the regulations severely threatened the economy of their operations. Finally as OSM's enforcement chief, Director Hall, noted, apparatus was designed to preclude the possibility of Appalachian operators continuing their old practices (personal interview, July 29, 1980). Small and medium-sized eastern operators based their preferences on this assessment. They recognized SMCRA's likely impact and formed their preferences accordingly.

Although as a rule the preferences of eastern firms imply implacable opposition to SMCRA, the virulence of that opposition varies within the region. Specifically, the most adamant and outspoken opponents of SMCRA were the smallest firms surface mining coal in the most mountainous region. This difference within the eastern region is interesting for two reasons. First, it indicates varying preference intensities, which

Table 5.1   Reclamation costs for a "typical" contour mine in the East

Type of reclamation	Incremental cost/ton
None	$0.00
Shape spoil bank	0.39
Terrace backfill	0.69
Contour backfill	0.95
Modified block cut	0.56
*Augering/narrow bench	0.00
*Augering/backfill	0.06

*Auger mining ceased after 1975.
Source: *CEQ Study*, 1973b.

suggests, in turn, an explanation for different behaviors within regions. Firms with the same preference orderings may be more or less active depending on how intensely they hold their preferences. One explanation for small firms holding the most intense preferences is that, for the most part, such coal firms in the East are directly owned and operated by one individual, frequently continuing a family tradition. Second, intraregional differences reinforce the importance of geography as an explanatory factor in firm behavior. As we shall see, some firms have a higher preference intensity because their location means that SMCRA has a more severe effect on them than most other similarly sized firms in the same region.

The most vehement and active of the small firms surface mined coal in Virginia's seven southwestern counties of Buchanan, Dickinson, Lee, Russell, Scott, Tazewell and Wise. Indeed, surface operators from Lee, Tazewell and Wise counties had formed the Tri-County Independent Coal Operators, a trade association that fought SMCRA vigorously at every turn. Several of these Virginia operators stated their views on SMCRA during congressional hearings. One such operator, Franklin Delano Robertson, testified: "In order for Virginians to surface mine Virginia coal under this bill [H.R.2], the steep slope provision, the return to approximate original contour provision, the data gathering provisions, the permitting provisions, the bonding provisions and citizen suit provisions must all be modified" (House Hearings, 1977b, pp. 14–15). Coming in 1977 after more than six years of wrangling and compromises, these comments suggest nothing less than a wholesale revision of the bill, rewriting its most critical provisions as a first step. This, in some small measure, demonstrates the irreconcilability of small eastern firms.

Echoing this view, James McGlothlin, president of another southwest Virginia firm and a director of NICOA, on a Senate version (S.7), maintained: "the bureaucracy of acquiring a permit will totally eliminate small and medium sized operators from continuing in the surface mining business. Most of these operators do not have registered or certified engineers, hydrologists, the equipment or technicians to do core drilling, or most of the expertise required to obtain a surface permit under the Act and to continue compliance therewith. . . . The same operators are not substantial enough to acquire proper bonding" (Senate Hearings, 1977c, p. 303). This bleak outlook received support from an affadavit sent to Senator Lee Metcalf by Daniel Gregory, vice president of the First State Bank of Wise, Virginia. Gregory argued: "it is our opinion as area bankers that the costs of additional equipment to handle spoil placement, or providing for high-wall and bench covering and securing of the necessary engineering and geological services would be more than the smaller operators could bear. . . . We conservatively estimate that from 80 percent ot 95 percent of the surface coal mines in Southwest Virginia will be forced to close" (Senate Hearings, 1977c, p. 1144).

R. L. Wingfield, president of the Central Coal Company, flatly stated to Metcalf's subcommittee, "some coal operators have testified that if you could just give them a few crumbs, minor amendments, they could continue to mine coal. This is simply not the case." In these comments, Wingfield refers to previous testimony by a panel of Kentucky operators whose position it was that they would accept a federal surface mine law if it included substantial amendments along the lines suggested by Franklin Robertson. The Kentuckians, incidentally, had strenuously opposed H.R.2 (Senate Hearings, 1977c, p. 306).

Summing up the views of the Virginia operators, B. V. Cooper, executive director of the Virginia Surface Mine and Reclamation Association, stated to Morris Udall's Subcommittee on Energy and the Environment that SMCRA "is not a reclamation bill. It is purely and simply a land use bill and a ban strip mining bill. The word 'ban' is never used, but the result is just exactly that—a ban. . . . There are literally dozens of impractical, shortsighted, punitive, harassing, and otherwise unreasonable requirements in this bill" (House Hearings, 1977b, p. 360). The strident tone and tenor of the Virginians' publicly stated preferences must be attributed substantially to firm size and the mountainous terrain of southwest Virginia. Limited capital, technical expertise and access to financing all serve to prevent such firms from adapting to environmental regulation. In the seven coal producing counties of southwest Virginia, as of 1977, 517 surface mines produced 13 million tons of coal. That, evenly divided, works out to approximately 25,000 tons annual production per mine. These production figures, in fact, range between 7,500 and 150,000 tons per mine yearly. With over 90 percent of these surface operations on slopes of 20 degrees or more, opposition to SMCRA was virtually assured.

If less vocal in their opposition, other eastern firms, both small and medium-sized, did not diverge markedly from the Virginians in the substance of their preferences. John L. Kilcullen, Washington counsel for NICOA, contended in congressional testimony that "it is the small and medium sized operator that has felt most severely the impact of policies and practices of the Federal Government" (House Hearings, 1973c, p. 130). Kilcullen, moreover, characterizes these firms as "100 percent totally uniform on SMCRA" (personal interview, February 2, 1981). Though Kilcullen was engaging in hyperbole—to be expected from a trade association official—his view accords with those of environmentalists and eastern operators mining outside Virginia. For example, two engineers for Kenvirons, an environmental consulting firm serving Kentucky surface operators on reclamation matters, testified: "It is an unfortunate fact of increased regulation that in and of itself it begets bigness. The small independent businessman in coal . . . simply cannot cope with the complexities of such increased regulatory requirements and gives up in favor of the larger operator who has the staff and/or capital to employ specialists capable of generating the data with which to comply" (Senate Hearings, 1977c, p. 90).

For their part, representatives of the Citizens Coal Project expressed no doubt about the negative preferences of eastern firms in general. EPI saw these firms as recalcitrants, unwilling to even entertain the prospect of federal regulation (House Hearings, 1977b, v. 3). The National Coal Association also confirmed Kilcullen's view of eastern preferences. According to Raymond Peck, chief counsel for NCA, "Even though big firms were adamant and persistent in their objections to SMCRA, there is a perceptible difference from smaller eastern outfits" (personal interview, June 25, 1980). Peck noted that eastern firms tended to be "more hysterical because their survival was at stake." Other trade associations supported this view. For example, Ben E. Lusk, president of the West Virginia Surface Mining and Reclamation Association, testified: "One of the big concerns of our association which represents a lot of smaller operators, is the discriminatory aspects of H.R.2 which we have noticed.

There is no possible way that a small operator can economically comply with the various permit requirements of H.R.2" (House Hearings, 1977b, p. 165). The fact that Lusk's organization supported SMCRA in principle (to eliminate competitive advantages for states with lax reclamation laws) makes this complaint more credible.

Larger firms or firms from other regions also recognized the untenable position in which SMCRA had placed their smaller eastern competitors. A senior vice president for operations at a large Midwest firm managing four underground mines and two surface mines producing a total of 6.2 million tons of coal per year noted that costs of surface mining had increased 20–40 percent as a result of SMCRA. He had no doubt that this affected smaller eastern firms most severely and was personally aware of three cases of mine shutdowns attributable to SMCRA in Appalachia (personal interview, July 28, 1981). Even segments of the eastern utility industry still dependent on coal for generating electricity added its voice to the general assessment of SMCRA's likely impact. In testimony before the Senate, B. B. Barker, president of the Duke Power Company, argued: "there is little doubt that H.R.2 will doom most independent operators in Central Appalachia. Recovery of coal under such restrictions (i.e., SMCRA), if it can be done at all, will have to be by large energy companies which can minister the required capital for the necessary massive investment" (Senate Hearings, 1977c, p. 247). (Naturally Duke Power preferred to purchase at least part of its coal from many smaller coal firms.)

In sum, there seemed to be little disagreement from any quarter, government, trade associations, utilities or other coal firms about SMCRA's likely effect on small and medium-sized eastern operators. The preferences of eastern firms reflected SMCRA's impact, differing only in the intensity of opposition.

If anything, the view of eastern firms hardened after 1977 once OSM began its Interim Regulatory Program. More and more eastern firms began to sound like the Virginians had in the legislative phase. This change is especially significant among West Virginia firms since several of them had supported SMCRA in the hope of easing competitive pressures from states that were like Virginia but which did not regulate surface mining as strictly. During 1979 Oversight Hearings on the Interim Regulatory Program, Thomas N. McJunkin of Amherst Coal Company, West Virginia, complained to a Udall subcommittee that "I just wanted to note that our basic approach today is not to criticize the Act. We feel, as I have tried to emphasize, that the regulations do exceed the Act" (Senate Hearings, 1979, p. 101).

Even more interesting is the testimony of Ben Lusk, director of MARC. Before founding this national trade association, Lusk had supported the concept of SMCRA to West Virginia operators and to Congress. By 1979, however, he could state: "what has happened since the passage of this Act is a crime. . . . The States are becoming more and more upset with the situation which has prompted many state governors to openly criticize the program. Many members of Congress supported P.L. 95-87 under false pretenses—that the highly technical regulations and OSM's overzealous enforcement isn't nearly what they intended" (Senate Hearings, 1979, p. 397).

In 1977 he had testified on behalf of West Virginia surface miners: "We believe that uniform guidelines are necessary to, one, bring about environmental improvement of land disturbed during the surface mining process; two, eliminate the uncertainty surrounding the industry . . . and three, to provide a more economically stable industry by the elimination of competitive inequities that are associated with the various differences in individual state regulation" (House Hearings, 1977b, p. 164).

Regarding the effect of the regulations and implementation, a senior vice president for a firm with eastern operations extracting approximately 2 million tons of coal per year from two surface mines said that he "didn't think the bill was that bad . . . though there were objections to specific provisions . . . when the regulation got through though there were 10,000 pages of regulations strictly interpreted" (personal interview, January 28, 1981). A consensus seemed to develop that the Interim Program went much further than the law intended, that the congressional staffers and environmental advocates running OSM tried to negate all the compromises embodied in the law by writing detailed regulations and strictly enforcing them.

As Joseph Krevac, director of water standards and criteria at EPA, an agency that worked closely with OSM on the Interim Program, noted, "The first people to staff EPA were shock troops committed to stringent regulation . . . but those on board for the long haul learned to go along to get along . . . calmer heads always prevail . . . OSM is in the shock troop stage now" (personal interview, July 22, 1980).

In the implementation phase of SMCRA, what sentiment there was for federal regulation among eastern firms evaporated almost completely. Eastern firms' preferences became for all intents and purposes, as Kilcullen suggested about the legislative phase, uniform. This consensus,

while cutting across categories of coal firms and including former supporters, did not encompass environmentalists, many congressmen and
regulators who felt OSM was too lax in enforcing its regulations and too
timid in dealing with state governments in the East.

The one complication in this otherwise straightforward expectation
about the preference functions of eastern firms is the fact that a number
of medium-sized firms in the East are captives. As such, they do not
have to cope with the problems of uncertain markets and extremely
difficult credit terms that confront other similar sized eastern firms.
Therefore, the possibility exists that medium-sized captives may have a
preference function that includes some amended version of SMCRA.
Particularly if these firms have planned expansion to meet the needs of a
parent company, the overriding economic interest, regardless of the
captive's individual profit function, is supplying requisite coal to the
parent's operations. Some captives, with the parent's backing, might be
willing to reach compromises on some issues to insure a more certain
environment for parent's coal demand.

In assessing SMCRA, a vice president for corporate development at a
captive firm pointed to problems of subsidence, detailed design criteria
and demands on management time as his chief worries. While he objected to "the extent of Washington's intrusion," he felt his firm could
live with SMCRA (personal interview, January 6, 1981). The manager for
surface mining operations at another captive bluntly stated that SMCRA
had no effect on his firm's level of production (personal interview, January 12, 1981). While this difference with the captive firms must be
noted, it serves as an exception that proves the rule. Although according
to 1975 figures 67 percent of all captive firms mined in the East, the
actual number was only twenty-six. All must be classified as medium-
to-large-sized firms. By 1979, the number of eastern captives had
dropped to twenty-two and their percentage to 60 percent of all captive
firms. The percentage drop owed to the influx of western firms run by
utilities.

In the implementation phase, even captives became less tolerant of
SMCRA, at least as administered by OSM. A major cause of their increased disenchantment lay in the unanticipated stringency of regulations dealing with the surface effects of underground mining. A reason
for the earlier sanguine view held by eastern captives had been that, as a
rule, 50 percent or more of their production came from underground
operations. One firm with 100 percent underground operations found
that it had to deal extensively with OSM on the regulation of refuse

disposal, mine road construction and nonpoint source pollution of streams and aquifers (personal interview, January 6, 1981). Most captives had had no idea of SMCRA's potential impact on deep mining.

The Midwestern Firms

In the Midwest, medium and small coal companies operate coal mines along with a substantial number of large operators. Mines as well as firms tend to be larger in the Midwest than in the East. This tendency is even more pronounced if we exclude southern Ohio and southern Indiana from the region—both have more in common with Appalachia than the Midwest. Together, these two areas account for 231 of a total 258 small surface mines and 54 of 72 medium-sized mines in the region. Nevertheless, even with all of Ohio and Indiana, midwestern coal mining encompasses fewer firms, larger firms and more surface mining than the East. In addition, regional characteristics differ significantly.

The small Ohio and Indiana mines in mountainous terrain notwithstanding, midwestern coal firms exhibited considerable flexibility in their preferences toward SMCRA. Even in the implementation phase, their increased opposition, directed largely against OSM, did not rival the East. In the midwestern region, firms, especially large firms, developed lexicographic preference functions based on certain provisions in SMCRA peculiarly relevant to their operations.

The flexibility of midwestern preferences stemmed, in part, from the fact that proponents of SMCRA took a far less adamant position on those portions of the program directly affecting the Midwest. Indeed, environmentalists' overwhelming concerns with restoring surface-mined land in the East and forestalling ecological calamity in the West diverted their energy if not their attention from the Midwest. This attitude is obvious from the legislative history of SMCRA. Not a single midwestern firm testified on H.R.2 or S.7, the two bills from which SMCRA was forged.

The overriding concerns in the midwestern region related to the vast amount of rich farmland in the area. Specifically, midwestern farmers and environmentalists tried to ban surface mining on "prime farmlands" and to insure that SMCRA would mandate returning surface-mined land to its original level of productivity. The mayor of a farming community in Illinois, for example, argued before a House subcommittee: "Although there is presently no strip mining in the Cutlin Township, Vermilion County has been the scene of much strip mining in the past, and the land still bears the scars from it. . . . Our most productive agricultural land should be preserved for raising crops and protected from strip mining. To this end, section 522 (a-3) should be amended to provide for the designation of prime agricultural land as not suitable for surface coal mining operations" (House Hearings, 1977b, p. 78, v. 4). Earlier versions of SMCRA vetoed by President Ford in fact placed a simple ban on all prime farmland surface mining. The final version merely empowered the Secretary of Interior to declare prime farmlands unsuitable for surface mining as he saw fit. It mandated only that surface-mined land be returned to its original level of productivity.

While prime farmlands provisions presented obstacles for midwestern surface operators, they did not invite the unyielding opposition engendered in the East by steep slope prohibitions and approximate original contour requirements. Also, the less vehement views of environmentalists toward midwestern firms raised the possibility of compromise. As the severity of these provisions declined from the total ban in S.425 to the more flexible approach of H.R.2, the preference functions of midwestern firms adjusted accordingly.

The potential impact of prime farmlands provisions varied considerably between large midwestern firms and small or medium-sized operators in the region. Large firms emphasized these provisions in a lexicographic form. In other words, they would consider amended versions of SMCRA if and only if a ban on prime farmland was weakened or eliminated. Since no midwestern firm participated in the public debate on SMCRA, this preference function is difficult to document. Nevertheless, it is possible to demonstrate it through a logical, if indirect, argument.

Because large firms have more resources and because they usually retain long-term contracts with steam electric utilities, they operate under a much longer time horizon than small and medium-sized firms. It also is easier for them to absorb or pass along increased regulatory costs. However, to fulfill existing contracts and to acquire new ones, these large firms need access to coal resources. Bans or restrictions on surface mining coal that lay on midwestern prime farmlands would limit that access. This would be the case even under H.R.2 if an "uncooperative" person were heading OSM or DOI since it would be his or her responsibility to designate lands unsuitable for surface mining. The whole question of bans and restrictions created a situation of uncertainty inimical to large firms' attempts at long-term development.

While this emphasis on problems related to prime farmlands provisions demonstrates, in a very specific way, large midwestern firms' desire for certainty, a much broader interest in certainty set them off from smaller firms. Citing "the attitude of people at osm" and smcra's "complex permitting process," an official of a large midwestern firm expressed concern about licensing areas for surface mining as needed to meet contracts (personal interview, December 20, 1980). Such concerns lend credence to the assertion that large firms in the region placed an extremely high priority on mitigating the impact of prime farmlands provisions on access to coal. Once access was assured, these firms could afford to consider other problems.

For small and medium-sized midwestern firms, prime farmlands presented a dual problem. If, as farm and environmental interests wanted, prime farmlands were to be designated as unsuitable, a great amount of strippable coal reserves would be withdrawn from commercial access. For these firms, this would mean a loss of opportunity to surface mine, the type of operation best suited to their size and financial capacity. However, even with the less stringent provisions contained in h.r.2, the costs of restoring the land to its original productivity posed just as serious a problem for small and medium-sized firms. In a report to the American Institute of Mining Engineers entitled "Meeting the Requirements of smcra," Peabody Coal Company experts estimated the incremental costs of prime farmlands compliance at between 16¢ and $4.00 per ton depending on local problems of topsoil handling and revegetation. For an interregional the size of Peabody, the problem was one of assuming these costs and then passing them on to public utility customers. For small and medium-sized firms, such incremental costs could spell the difference between profit and loss.

In general, the incremental costs and front-end costs of smcra posed a serious threat to smaller firms in the Midwest. Although area mining techniques and thicker coal seams allowed lower unit costs and higher productivity among these firms than their eastern counterparts, compliance costs were still significant in the Midwest. Moreover, market conditions made it difficult to finance these added costs. Kenneth Summers, an analyst writing in the *Journal of Commercial Bank Lending*, reported: "owing to the large capital requirements for opening and running a mine . . . the trend toward fewer but larger companies will continue coupled with the increased costs of production due to regulation and the inability of smaller companies to obtain necessary equity funds" (1978, p. 10). A year later, in the same journal, Thomas J. Tucker observed: "The requirements for permits, plans and other procedures under the regulation of smcra . . . pose a complication for many small operators. They simply do not have the administrative strength to cope with the paperwork and skilled technical analysis demands of the regulations" (Tucker, 1979, p. 33). While the particular provisions and regulations relating to prime farmlands uniquely affect central region firms, the numerous costly aspects of SMCRA—permitting, bonding, topsoil management, fugitive dust control and hydrological data requirements—all heavily influenced the preferences of small and medium-sized midwestern firms.

Once it became clear that Congress would not seriously consider legislation to phase out surface mining (i.e., the 1971 Hechler bill), small and medium-sized firms in the Midwest focused their attention on specific provisions in an attempt to gain amendments that would ease compliance. Analogously to their eastern counterparts, these midwestern firms concerned themselves with those provisions of SMCRA that would impose the highest costs.

To get an idea of the actual costs involved, we can turn to a study of SMCRA's impacts by Consolidation Coal. This previously cited study sought to contrast the compliance costs for SMCRA with what Consol termed "good engineering practice," that is, the environmental measures it had implemented or was planning to implement. Focusing on twenty-one specific requirements of SMCRA, Consol developed the following average cost estimate for its midwestern operations (see table 5.2). The crucial point from table 5.2 is that even partial compliance with SMCRA is quite costly. A large firm like Consol stood ready to assume substantial regulatory costs and had done so prior to SMCRA's enactment. However, for smaller firms even the costs of so-called good engineering practices were extremely burdensome. And as financial analysts indicated, adequate financing was not forthcoming for any except the large firms. The only thing mitigating these effects in the Midwest were the economies of scale associated with the region's surface mines.

After SMCRA's passage midwestern concerns focused on compliance costs: the prime farmlands provisions no longer posed as serious a threat to access. During Senate Oversight Hearings on the initial regulations, Neal S. Tostenson, executive vice president of the Ohio Mining and Reclamation Association, argued: "As a whole, the actual provisions relative to final reclamation will not differ in Ohio under the Federal program. We are already achieving high standard of reclamation . . . however, under this Federal program of regulations, the costs of coal will probably increase over 20%" (Senate Hearings, 1978d, p.

Table 5.2 Average cost estimate for Consol midwestern operations

Provision	SMCRA costs $/ton	"GEP" costs $/ton
Backfilling and grading	1.96	1.913
Blasting	.02	0.00
Coal processing plants	.01	.0076
Dam design and construction	.06	.04
Exploration standards	.0007	.00
Fugitive dust control	.42	.11
Hydrology standards	.02	.002
Mulching	.03	.02
Permitting	.07	.02
Prime farmlands	.32	.00
Reconstruction of mines	.09	.01
Revegetation	.17	.09
Road construction	.16	.09
Sedimentation ponds	.098	.05
Soil disposal—deep mines	.014	.00
Subsidence control	1.66	.00
Topsoil standards	.96	.73
Total cost	6.0327	3.0826

Source: Consol, 1980.

128). But Tostenson complained bitterly about the fact that OSM regulations would require, within a three-month period, that all existing structures on mine sited be modified to comply with SMCRA's regulatory standards. In practice, OSM could regulate the effects on water and soil pollution of deep mines' surface activities, a major concern for large and medium-sized firms. For the Midwest's deep mines, this would include preparation plants, roads and rail spurs, all of which would demand heavy and immediate capital investment from firms regardless of their size.

Specifically addressing the situation of small and medium-sized midwestern firms, Tostenson concluded: "throughout the regulations we have mentioned . . . the regulations were written to cover all situations regardless of size and financial ability of the operator to cover these costs" (Senate Hearings, 1978d, p. 128). This was a major concern in southern Indiana and southeastern Ohio where hundreds of small independent surface mines operated. Facing more rugged terrain than most midwestern firms, these operators shared some of the compliance concerns of steep slope Appalachian firms. However, even for the medium-sized midwestern coal firms, SMCRA's costly provisions aroused objections that played a part in their preference functions.

The major contrast in preferences among midwestern operators was between large firms and medium or small firms. A vice president for public affairs at a large midwestern coal company maintained that, with regard to SMCRA, a preference for certainty separated major firms in the region from others (personal interview, January 5, 1981). The practical meaning of this statement was that major midwestern firms were willing (and could afford) to acquiesce to costly provisions when smaller firms would have preferred "to tough it out with the environmentalists" (ibid.). The larger firms would cooperate with pro-SMCRA forces to an extent in order to get a surface mining law on the books and get on with the business of mining coal, provided they could have access to coal. This formed the basis of a quid pro quo with environmentalists that was beyond the reach of smaller firms. The costly provisions covering mining operations that were of such moment to small and medium-sized firms were of far less import to large firms in the central region. Permitting and bonding procedures worried larger firms only insofar as they imposed delays on mine development.

The foregoing remarks should not imply that large firms were indifferent to regulatory costs attributable to SMCRA. These firms simply felt capable of assimilating many of the costs over time. They sought to minimize the costs as a secondary aim. According to the senior vice president of operations for a midwestern firm producing 7 million tons of coal per year, approximately one-third of it by surface mining, "SMCRA raised surface mining costs 25–40 percent depending on the characteristics of the individual mine" (personal interview, January 28, 1981). Clearly such cost effects were cause for concern, though more so in the implementation phase and for smaller firms.

The Western Firms

As indicated by the description of the coal industry, the nature of surface mining in the western region precludes all but large firms. The annual production of a single western mine usually exceeds that of an entire large midwestern firm. Firm size and experience with detailed state surface mining regulations in the West provide firms in that region with the capacity, if not the predisposition, for compliance with a federal law. The preference functions of western firms reflected these factors.

In addition, western coal mining, being a relatively new segment of the industry, was not burdened with coal's "bad boy image" among pro-SMCRA forces. Western firms thus stood in a better position to reach compromises on SMCRA than other regions. This situation was partially offset by environmentalists' concern over the prospects of ecological damage associated with the projected boom in western coal surface mining. Pro-SMCRA forces were not prepared to make wholesale accommodations on certain peculiarly western provisions in SMCRA, though none of these provisions posed an insurmountable obstacle to the surface mining of western coal.

As was the case with large midwestern firms, the provisions requiring substantial expenditures for compliance caused far less concern than those provisions and regulations that created uncertainty or posed a risk to access to coal reserves. Large western firms were most concerned not over those aspects of SMCRA that impose incremental regulatory costs, but rather those that menaced plans for coal mine expansion or development. This is an especially crucial consideration in the West where billions of investment dollars were tied up in coal projects still in the planning stage when SMCRA came on the national agenda.

Two provisions, one dealing with surface owner consent and the other with surface mining on alluvial valley floors, uniquely impinged on western firms in this regard. Both sought prohibitions on coal surface mining, which would have affected access to rich western coal fields. Western coal firms concentrated, therefore, on mitigating the effects of these two provisions and their companion regulations. Succeeding elements of their preference functions related to those aspects of SMCRA directly responsible for increasing mining costs. In other words, western firms also exhibited lexicographic preference functions. They placed primary importance on mitigating problems of uncertainty and access raised by alluvial valley floor prohibitions and surface owner consent.

Western firms' preference functions were further complicated by the importance assigned to SMCRA's provisions dealing with citizen participation, citizen suits and a public role in permitting. These provisions also raised the prospects of ambiguity, delays and limited access to coal. Therefore, they constituted a secondary level in western firms' lexicographic preference function: after victory or defeat in mitigating alluvial valley and surface owner consent provisions, western firms would consider amendments to SMCRA if and only if citizen participation were weakened.

A major fear of western firms was the so-called Mansfield Amendment introduced by the senior senator from Montana. Introduced in the 93rd and again in the 95th Congress, this amendment would have banned all surface mining of coal on private land though the resource is legally public property. Even with the amendment's eventual defeat, SMCRA provided that private surface owners must consent to stripping operations. The problem for western operators was not so much the amount of coal that these provisions would withdraw from commercial development as the fact that with checkerboard ownership patterns in the West, one hidebound rancher or concerned environmentalist could jeopardize the economy of a multimillion-dollar coal mine.

That environmentalists placed a high priority on surface owner consent was clear. For example, Gerald Moravek of the Powder River Basin Resource Council demanded "the inclusion of strong surface owner consent provisions" (House Hearings, 1977b, v. 4, p. 224). These provisions, in the eyes of western firms, raised the possibility of limited access and great uncertainty given the vagueness of the language employed. One thing was clear, however: the provisions would vest enormous power in the hands of landowners of the Northern Great Plains and the Southwest, many of whom were environmental activists themselves or sympathetic to their cause. Speaking for ARCO, H. E. Bond noted: "Landowners in the West . . . acquired title to surface lands fully realizing that they did not control the vast natural resources of coal . . . lying beneath . . . under this Bill and contrary to existing law, these same landowners could prohibit the development of coal" (House Hearings, 1977b, v. 3, p. 274).

A single landowner had the potential to disrupt the development of an entire mining operation if his landholding was crucial to the economy of that operation. Even if landowners could not make their "vetoes" of surface mining operations hold up through an appeals process, the length of that process could seriously delay mining or jeopardize previously arranged financing agreements. Articulating the typical viewpoint, the president of a firm surface mining in North Dakota and Montana expressed exasperation over his experience with surface owner consent providing environmentalists "a ready veto" and unscrupulous individuals with a means of "extortion by surface ownership" (personal interview, February 5, 1981). For western firms this presented a far more serious challenge than meeting the incremental costs of SMCRA.

SMCRA's provisions on alluvial valley floors evoked an equal level of concern among western firms. The two sets of provisions were emphasized in the congressional testimony of every western firm before every subcommittee between 1972 and 1977. Significantly, most of this testimony was not orchestrated by a trade association—firms from other regions usually appeared as part of a panel organized by the National Coal Association, American Mining Congress, National Independent Coal Operators Association or some lower level group.

This independence of western firms reflected a number of factors. First, these firms were large enough and had sufficient political experience that trade associations did not provide significant economies of scale. Second, and most important, western firms saw association with other segments of the industry (through NCA or AMC) as a political liability. By operating independently in regulatory politics, western firms appeared less threatening and more responsible than the traditional coal industry.

Turning to the second major provision affecting the West, there is no doubt that environmentalists thought alluvial valley floor provisions were important. Wallace McCrae of the Northern Great Plains Resource Council, for example, argued that "the provision protecting alluvial valley floors is of primary importance. . . . I cannot emphasize too strongly that alluvial valley floors have to be protected from surface mining" (House Hearings, 1977b, v. 4, p. 231). The import of this provision certainly was not lost on western firms that, without exception, in the 93rd and 95th Congress hearings on SMCRA, highlighted alluvial valley floors as a provision to which they strenuously objected. In typical testimony, H. E. Bond of ARCO's minerals division, operating mines in Wyoming's Powder River Basin, stated that there was "extremely ambiguous language contained in the alluvial valley floor provision" (House Hearings, 1977b, v. 3, p. 270). The problems, again, were that SMCRA injected considerable uncertainty into investment planning and jeopardized coal mine development with the threat of removing a small but crucial adjacent landholding from commercial exploitation.

In the legislative phase, western firms also criticized the bill's provisions on public participation in the permitting process and citizen suits. Western firms placed a high priority on these provisions because of their potential for uncertainty and disruption. Past legal actions of environmentalist groups in the West no doubt served to kindle such fears. In particular, *Sierra Club v. Morton* had forced DOI to suspend coal leasing in the early 1970s. The merits of that decision notwithstanding, the prospect of environmental "extremists" continually seeking injunctions through the citizen's suit mechanism or delaying operations through challenges in the permitting process worried western operators. In addition, they felt this would be unfair considering their relatively good record at compliance with state and federal regulations all along.

Given the massive capital investments tied up in western mines, even a slight possibility of an injunction or other delay could cause concern. T. L. Austin, chief executive officer of Texas Utilities Company, which operated lignite mines in Texas, remarked sarcastically, "This is not a reclamation bill; it is a guaranteed relief and retirement act for activist lawyers" (House Hearings, 1977b, v. 4, p. 254). In conjunction with these fears, western firms also indicated uneasiness at the prospect of court actions based on the unclear use of terminology in the bill. ARCO argued in both the Senate and the House that "more precise language is needed so as not to invite litigation with its attendant delays of coal mining projects. (House Hearings, 1977b, v. 3, p. 275).

It is important to realize that many western firms could accept provisions for citizen activism, but they feared abuses. For example, Kenneth Holum of Western Fuels testified: "We welcome review at the administrative level. We consider that opportunities for judicial review at the insistence of interested citizens is clearly appropriate" (Senate Hearings, 1977c, p. 667). Holum, however, noting the public position of FOE and other environmental organizations, felt that there was a high probability of abuse in some quarters. For western firms, the costs of citizen participation could be more of a burden than direct regulatory costs.

According to the head of the Government Relations and Reclamation Department (a revealing title) at a western firm, his company "produced a ton of paper for every ton of coal" under SMCRA. He estimated that whereas under state regulation he devoted 98 percent of his time to mining, he now spent 50 percent of his time on reclamation (personal interview, December 14, 1980). Even if this is an exaggeration, SMCRA, through its provisions on participation, placed significant new demands on management. Another western firm executive suggested that "90 percent of the problem with environmental influence at OSM is no experience in mining and they don't understand state peculiarities" (personal interview, February 9, 1981). Less charitable western firms suspected that environmentalists influential at OSM deliberately sought to use SMCRA to harass western surface mining. These complaints focused, perhaps predictably, on the field inspectors. The chief executive officer of a western firm that had supported federal regulation of surface mining felt that "the function of lower staff people at OSM, the GS 12–14 group, is to obstruct mining" (personal interview, January 30, 1981).

The president of another western firm characterized OSM inspectors as "gestapo agents whose self-appointed mission, it seemed, was to harass surface mine operators rather than follow the intent of the law" (personal interview, February 5, 1981). This view among western firms emerged only in the implementation phase in which many operators felt that environmentalists who helped write the regulations pushed OSM beyond the intent of the law.

In conclusion, the western region is composed of a homogeneous set of firms. All western surface mining began no earlier than the mid-1960s. In addition, western firms, from the outset, gained valuable experience in dealing with extensive state reclamation programs and federal bureaucrats. Not only were the firms themselves large, but individual mines, the basic regulatory target, were also gigantic in comparison with mines in other regions. All of these factors evoked a preference function that emphasized concern over access to coal and reduction of risk. Secondarily, western firms preferred legislation that would minimize the impact of those provisions directly regulating mining activities. Concern with these, though, also centered on the prospects for delay and obstruction rather than incremental compliance costs.

The Interregional Firms

Although interregional firms are, almost without exception, large, assessing their preferences toward SMCRA is problematic. On the one hand, all of the foregoing remarks about large firms' ability to afford compliance through internal funds or access to external financing and about in-house technical expertise apply to interregionals. On the other hand, precisely because their huge mining operations cut across regions their preference functions tend to be less clear than those of large midwestern or western firms. To the factors of size and region, then, we can add regional dispersion as an explanation of firm preferences (Salamon and Siegfried, 1977). Firms with dispersed operations will have more complicated preferences.

The reason for this complexity, of course, is that SMCRA affects the three regions differently and interregionals must comply for their operations in each region. Their preferences, naturally, will depend on the regions in which their operations predominate and the regions in which expansion is planned. This can create some vexing situations for the interregional firm. For example, SMCRA may offer certain competitive advantages to operations in one region at the expense of those in another. As we know, growth and increased productivity were forecast for western surface mining at the expense of the East and the Midwest.

To the extent that interregionals plan significant expansion in the West, it is an inescapable conclusion that SMCRA will favor their larger western mines over underground and surface mines in other regions. Expansion plans, thus, superimpose a set of preferences akin to those of western operators on basically eastern- or midwestern-oriented preferences of interregionals. In fact, it is convenient to speak of eastern-oriented or midwestern-oriented interregional firms, both of which have substantial interests in common with western firms. These western interests modulated their basically eastern or midwestern preferences.

Another complication in the preferences of interregionals stemmed from the fact that, in negotiations with the UMW, they depended on the support and alliance of medium-sized firms in the East and the Midwest. No matter what their pure preferences in relation to SMCRA, interregionals had to consider the costs of alienating smaller firms on which the costs of SMCRA had a more deleterious effect. While this did not impose an absolute constraint on the preferences or behavior of interregionals (especially in light of their growth in the nonunion western coal industry), it did complicate their interests.

The eastern-oriented interregionals actually have little in common with the small and medium-sized operators of that region. The large eastern operations are, for the most part, underground. Large surface mines are mountaintop removal and head-of-hollow fill operations or open pit mines adjacent to deep mines. Rarely do interregionals operate contour mines. As a result, the steep slope and return-to-approximate-original-contour provisions that figured so prominently in the calculations of other eastern firms did not pose as great a threat to interregionals.

Environmentalists viewed the surface operations of these firms with considerably less trepidation than those of smaller firms. Interregionals could afford to comply with SMCRA or employ technology which did not pose a severe ecological threat. With some opportunity for accommodation, therefore, eastern-oriented interregionals did not exhibit the dichotomous preference function of their eastern competitors. Rather, they concentrated on achieving compromises within SMCRA to facilitate and insure the continuation of "acceptable" mining techniques, while suggesting modifications in the stringency with which steep slope prohibitions were to be applied. As large firms, their interest lay in insuring access to coal deposits, not in alleviating incremental regulatory costs.

Another major concern about SMCRA for these firms was the law's provisions to regulate the surface effects of underground mining. Although this set of provisions remained vague in the legislative phase, the preference of eastern-oriented interregionals included it during the implementation phase when the nature and extent of the impact became clear. The director for corporate environmental quality control at an eastern-oriented interregional admitted that he had not envisioned "what a monumental headache underground effects would be" (personal interview, January 9, 1981).

Midwestern-oriented interregionals exhibited a similar, though slightly different, set of concerns about SMCRA. Like their eastern counterparts, midwestern interregionals were concerned about access to coal in their base region. Interregionals with operations based primarily in the East differed from those with operations based in the Midwest. In assessing SMCRA, the former feared that banning steep slope mining would impinge on their operations, while the latter expressed concern about the likely impact of bans on surface mining on prime farmlands. Eastern-oriented interregionals were eager to legitimize mountaintop removal and head-of-hollow fill. Midwestern-oriented firms focused on easing or eliminating a total ban on prime farmland mining.

Like their eastern counterparts, midwestern interregionals engaged in significant amounts of underground mining. Consequently, midwestern firms too were concerned about SMCRA's regulation of underground mining. However, midwestern firms were clearly weighted toward surface mining. For example, two eastern-based firms estimated their proportion of surface mining tonnage as 20 percent and 10 percent (personal interviews, October 30, 1980, and January 9, 1981). Two midwestern interregionals, by contrast, indicated that their operations included 70 percent and 95 percent surface-mined tons (personal interviews, January 9, 1981, and January 5, 1981). Since each of these four firms operated only surface mines in the West, and since their western operations accounted for comparable proportions of production, these figures reflect differences between eastern and midwestern operations. The eastern-based firms were far more interested in SMCRA's control over surface effects of underground mining. During interviews, the officials of the two eastern-based firms stressed the proportion of surface mining as a major factor in assessing SMCRA's impact. Neither of the two midwestern firms labeled the surface effects of underground mining as a major concern. In fact, one of these firms failed to mention it, even though the firm operated some large underground mines in Illinois. Ostensibly, these problems received a lower priority in the Midwest.

In spite of the differences between eastern and midwestern interregionals, both resemble western firms in several respects. Interregionals developed some of the first major western surface mines. Consequently, their experiences with state and federal regulation as well as their size and access to professional expertise matched that of western firms.

Similarities in preferences between interregionals and western firms during the legislative phase emerged in a common interest about SMCRA's potential for jeopardizing financial arrangements, delaying production and cutting off access to coal reserves. Like western firms, interregionals worried about citizen suits and public participation in the permitting process. In the words of Gene Samples, president of Consolidation Coal, an interregional headquartered in Pittsburgh: "The one aspect of this bill (H.R.2) which constitutes the single greatest impediment to increased coal production is the potential for endless and repetitive litigation inherent . . . in the bill" (Senate Hearings, 1977c, p. 391). Samples noted that the section of the proposed legislation which granted automatic standing for citizen suits against individual firms raised major fears among coal companies and their creditors. The combination of ambiguity in the bill's language and the ease of bringing suit could tie up a coal project for years.

While the financial resources available to interregionals were considerable, they were not inexhaustible. These kinds of delays could undermine the profitability of a venture. In a similar vein, the vice president for public affairs at a midwestern interregional underscored the length of time and potential for delays inherent in the permitting process (personal interview, January 5, 1981). In interviews, interregionals consistently stressed this aspect of the permitting process under SMCRA. When incremental compliance costs did come into discussions, it was in the context of their effect on planning and lead time necessary to develop a coal mine.

The second major area of concern among all interregionals in the legislative phase was the provisions that had the potential to remove substantial western coal deposits from commercial development. For western operations, those provisions dealing with alluvial valley floors and surface owner consent figured as prominently in the preferences of interregionals as of western firms. Since the planned investment of these firms focused on the West, these provisions were crucial.

In testimony before the House, a panel of executives from Consolidation Coal and AMAX, a midwestern interregional, emphasized their concerns about SMCRA's potential effects on their western operations. Samples of Consol argued: "too many of the provisions of this bill state or imply a simple prohibition of surface mining. . . . An example is the section dealing with surface owner consent . . . the bill instructs the Secretary of Interior . . . to the *maximum extent practicable* to refrain from leasing Federal coal deposits underlying privately-owned surface lands" (House Hearings, 1977b, p. 393). John H. Paul, representing AMAX, devoted six pages of testimony to a detailed discussion of his firm's concerns about the probable impact of alluvial valley floor provisions. The majority of these comments pertained to a "lack of clear definitions and varying interpretations of all applicable subsections" (House Hearings, 1977b, p. 403). Like Samples, Paul feared that strict interpretation of SMCRA would preclude access to western coal. Also, the ambiguity, they felt, would imperil financing for multimillion-dollar mines.

As SMCRA passed to the implementation phase, the primary attention of interregionals shifted from provisions that could restrict access or invite litigation to the permitting process itself. Again, their preferences paralleled those of western firms. Interregionals found themselves committing tremendous internal resources to acquiring new permits. Typically, a filing procedure would involve the production of a 2,500-page report with over two hundred maps (personal interview, January 5, 1981). These firms consistently remarked on the greatly expanded time needed to bring a mine into production. Each permit application must include legal, financial and environmental data and a detailed reclamation plan. The environmental data alone must be collected over the course of a year at a minimum. Altogether, the permitting process takes between 2.5 and 3 years (Jones, 1979b). Although interregionals had the capital and managerial resources to comply with permitting procedures, they sought to ease the burden on management time.

The positions taken by interregionals and, by extension, their preferences mirror those of western firms. SMCRA's provisions affecting risk and access dominate their considerations. These provisions include those pertaining to alluvial valley floors, surface owner consent, citizen suits and public participation in the permitting process. William Dickerson, EPA's director of Office of Environmental Review and an architect of SMCRA, characterized the concerns of western and interregional firms by noting that "They say, 'Tell us what to do' and they do it, unlike small firms concerned with the dollars to comply" (personal interview, July 21, 1980). This preference is consistent with the view of major firms from the earliest part of the legislative phase. Commenting in 1968 on the prospect of a federal surface mining law, the National Association of Manufacturers argued that "a perpetual, overhanging possibility of Federal intervention . . . would make realistic planning—from both operational and economic standpoints—practically impossible" (Senate Hearings, 1968, p. 307).

Owing to the complex nature of SMCRA and the dispersion of interregionals' operations, however, some obvious differences exist between the preferences of interregionals and western firms and among interregionals themselves. First, interregionals have a wider range of concerns than western firms. Interregionals operate deep mines as well as surface mines. They also mine coal in different regions whose geological, climatic and topographical features uniquely determine SMCRA's effects. In addition, the regional dispersion evokes a difference between eastern-oriented and midwestern-oriented interregionals. Eastern-based firms devoted a great deal of secondary attention to steep slope provisions and "acceptable" surface mining techniques. These same firms also feared the effects of regulating surface effects of deep mining. Midwestern firms, on the other hand, worried about prime farmland provisions in their second order preferences. They also worried about the regulation of deep mining, though to a proportionally lesser extent than eastern firms. Because of the complicated, even conflicting, preference considerations of interregional firms, the ordering of social outcomes and, ultimately, the firm's behavior can depend very heavily on the firm's leadership. The more complex the preferences based on economic and financial considerations, the more discretion will logically devolve on firm leadership.

Firm Leadership

Differences among large firms, particularly interregionals, emerged for which regional factors simply cannot account. Nor can size—all interregionals are sufficiently large to have similar interpretations of SMCRA's economic and financial impacts. The inescapable conclusion growing out of interviews and congressional testimony was that ideological and policy differences, often attributable directly to chief executive officers, explain individual differences among large firms. As noted with regard to smaller eastern firms, the acrimony of their opposition could vary with leadership. These differences count for more among larger firms because these firms' resources afford them the luxury of diverging on ideological or policy issues. Smaller firms usually have to deal with a more direct threat to their economic welfare from environmental regulation.

Peter Gabauer, a regulatory specialist at NCA who gained experience on congressional staffs and in the Commerce Department, pointed out that some coal firms simply oppose government intervention on ideological grounds and would not cooperate even to the extent of trying to make federal regulation more streamlined. He noted, somewhat cryptically, that in the chemical industry, du Pont historically had cooperated with "the Feds" while Dow steadfastly refused (personal interview, March 25, 1980). Supporting this position, Ray Peck of NCA suggested that variations in sophistication and willingness to deal with the bureaucracy depend largely on the chief executive officers at coal firms (personal interview, May 12, 1980). Both men were discussing interregionals and the problems NCA faced in coordinating political action among major firms in the trade association.

Although it is possible to trace the public positions of several large firms in an attempt to bring out the importance of leadership differences for regulatory preferences, the most compelling evidence comes from within firms. This owes to the fact that, under the aegis of NCA and AMC, interregional firms strove to delineate a common ground on SMCRA. This attempt at a "united front" masked many underlying discrepancies owing to firm leadership. In part, these also owe to the fact that major concerns with SMCRA are so similar that it would require very subjective judgments to identify ideological and policy differences.

A good case for the role of leadership emerges from changes within a firm over time. The chief counsel for an eastern-based interregional pointed out that his company's position on SMCRA paralleled "changes in personnel and attitude at the top" (personal interview, October 30, 1980). The earlier verions of SMCRA had evoked a vehement negative reaction. With a change in top management, the policy reportedly shifted toward "pragmatic accommodation." Although the political reality of SMCRA's likely enactment may have strengthened the hand of the pragmatic accommodator at this firm, there is no question about there having been a definite philosophical shift.

Leadership also made a difference between interregionals. In contrast to the example of pragmatic accommodation cited above, another large eastern interregional vigorously opposed the concept of federal regulation throughout the legislative phase. This second firm "moderated its position" after SMCRA's passage according to one firm official (personal interview, January 12, 1981). However, even in the implementation phase, this firm was in the forefront of the fight to weaken SMCRA and OSM. Similar ideological differences emerged between midwestern firms. In particular, the reclamation approach of Peabody differed markedly from that of AMAX. Norman Williams, assistant director of OSM, singled out Peabody as a responsible firm with a strong history of reclamation (personal interview, July 28, 1980). Dick Hall, OSM's director of enforcement, on the other hand, cited AMAX among a handful of major firms that was "sharply opposed" to federal regulation (personal interview, July 29, 1980). This difference in approach between similarly situated firms reflected a difference in management philosophy. Interestingly, this difference was more pronounced in the legislative phase than the implementation phase, unlike other interfirm differences. After SMCRA's enactment, AMAX and Peabody as well as the two eastern firms discussed above were confronted with the realities of compliance. This tended to submerge ideological or policy differences among them.

Finally, leadership differences cropped up among eastern captive firms too. This makes sense insofar as their relationship to parent firms affords them a similar latitude in responding to SMCRA as that of inter-regionals. Norman Williams of OSM pointed out, for example, that some captive firms associated with steel companies "were inclined to go along to get along" when it came to SMCRA (personal interview, July 29, 1980). Moreover, he noted that as a congressional staffer in 1975, he was aware of an instance in which a steel-related firm had drafted a letter in support of overriding the second Ford veto. Only the intervention of Carl Bagge, president of NCA, persuaded the firm not to publicize the letter. E. P. Leach of Bethlehem Steel's mining division clearly exemplified this conciliatory approach toward SMCRA. In testimony before the House Subcommittee on Energy and the Environment, he stated: "Despite such higher costs [of reclamation in the East due to SMCRA], there are ways to accomplish reasonable reclamation. We agree not only that we cannot alter the landscape in an unacceptable manner, but that reasonable regulations should be enacted and enforced" (House Hearings, 1977a, v. 2, p. 369). In contrast, the vice president for coal operations at another steel captive reacted with outrage at the very idea of SMCRA. In discussing the early Hechler bill, he commented, "The only thing I'd like to discuss about Ken Hechler is his funeral" (personal interview, January 23, 1981). The tone and substance of the positions of these two firms differed markedly, and cannot be explained in any terms other than different management philosophies.

Table 5.3   Preferences on SMCRA by firm type

Size	East	Midwest	West	Interregional
		Region		
Large	Mild opposition to the regulatory program. Strong interest in allowing for cost effective stripping technology (e.g. mountaintop removal) and restricting public participation.	Mild opposition to the regulatory program. Strong interest in eliminating the ban on prime farmlands as well as public participation provisions.	Strong opposition to alluvial valley bans and public participation. Interest in demonstrating "social responsibility," contrasting performance with "old" firms and getting a law on the books as soon as possible.	Mild opposition based on resentment of government intervention. Specific interest in eliminating all bans as well as public participation. Strong interest in resolving uncertainty.
Medium	Strong opposition to the regulatory program based on inability to meet compliance costs and resentment of government intervention.	Opposition to the regulatory program based on significant impact of compliance costs. Strong opposition to particular provisions (e.g., prime farmlands).		
Small	Strong opposition to the regulatory program based on inability to meet compliance costs and resentment of government intervention.	Strong opposition to the regulatory program based on inability to meet compliance costs and resentment of government intervention.		

Conclusion

It seems clear that SMCRA's attempt to address the salient environmental problems associated with surface mining called forth different responses from different categories of firms. Both region and size played significant roles in determining firm preferences. However, probing a bit deeper reveals that another factor, firm leadership, also can account for important differences among coal firms. Leadership refers to the individuals and/or parent companies responsible for setting firm policy.

In categorizing firms, this type of information was omitted except for noting that captive firms might diverge from similarly situated non-captive coal operators. It was assumed that these leadership differences would parallel region and size differences. The investigation, however, reveals that leadership has an independent effect on preferences. Leadership differences can account for what otherwise would be inexplicable interfirm differences, differences with no apparent basis in firm profit functions.

The idiosyncratic aspect of firm leadership notwithstanding, we can make some general statements about firm preferences. Specifically, firms did differ systematically in their preferences. These differences reflect fundamental, underlying economic variations that, in turn, depended on firm size and the location of surface mining operations. Table 5.3 illustrates these interfirm differences.

6 SMCRA and Firm Behavior

The Calculus of Participation

Given the variations in firm preferences within the coal industry, one should expect a corresponding variation in behaviors. Firm preferences, tempered by an assessment of political and economic realities, guide firm behavior in regulatory politics. In their attempt to relate their economic ends to political means, individual firms, or for our purposes categories of firms, may be said to employ a *rational calculus of participation*. By this I mean the decision-making procedure that underlies a firm's judgment as to whether and how to participate in regulatory politics.

A calculus of participation requires a firm to compare the value of a particular policy outcome with the costs of influencing regulatory politics to achieve that outcome. Since the likelihood of a firm's success is a function not only of political activity costs, but also the political-economic framework of regulatory policy, that framework also is an element in its calculus of participation. This chapter considers the various calculi of participation employed by individual firms.

The Eastern Firms

The political response of eastern coal firms reflects their vituperative opposition to SMCRA. From the standpoint of eastern firms, SMCRA left no ground for compromise. In view of the front-end costs, steep slope provisions, approximate original contour provisions and compliance costs, small and medium-sized operators in the East bore the brunt of

SMCRA's impacts. John L. Kilcullen, chief counsel for NICOA, asserted plainly that "Appalachia got a raw deal in comparison to the Plains [i.e., the Midwest] and the West" (personal interview, February 2, 1981). As we have seen, this perception was not limited to eastern firms themselves. All participants in the development of SMCRA readily acknowledged the stringency with which eastern surface mining was to be regulated, though there was a difference of opinion as to the justification for this. Ray Peck of NCA, for example, pointed out that SMCRA was a matter of "survival" for eastern firms (personal interview, June 15, 1980). William Dickerson of the EPA agreed but noted that for these firms "the only concern was dollars to comply, profits, and the environment be damned" (personal interview, July 21, 1980).

With rare exception, eastern firms' preferences reflected uncompromising opposition throughout the legislative and implementation phases of SMCRA. However, strong preferences alone are not sufficient to warrant political action by a rational firm. If no chance of success exists or if the costs of participation are extremely high, it may be prudent to give up the fight against SMCRA.

The political-economic framework confronting eastern firms presented a low probability of success for overturning SMCRA. On the plus side for eastern firms, the nation, in the mid-1970s, had begun to look toward coal as an energy alternative as OPEC prices escalated. Therefore, it was reasonable to argue that it was the wrong time to impose a regulatory program that would severely curtail or eliminate coal production from thousands of small and medium-sized producers. However, this political advantage was undercut by every other element of the political-economic framework. Indeed, such arguments had been unsuccessful in defeating the Clean Air Act. Perhaps most important, the Congress and the public looked to western coal fields as the basis of increased coal production. Moreover, low-sulfur western coal seemed to offer a good environmental alternative for protecting the modest gains achieved in air quality.

The focus on increased western coal production also received support from various western legislators, a number of whom, in 1972, took over key positions on the congressional committees responsible for SMCRA. These legislators and their environmentalist allies already had made concessions in limiting SMCRA to coal surface mining, reducing assessments for the orphaned mine fund and providing financial aid to small operators on data requirements. They were not about to relax performance standards, though, especially those bearing on Appalachia.

It was clear from the publication of the Udall Report in 1967 that surface mining's worst impacts were in the East, and the provisions of SMCRA bearing on that region would be the heart of the program. Environmentalists and key legislators wanted to preclude the possibility of a second Appalachia and to halt the mining practices that had despoiled the environment in that region. Conveniently for western legislators, this kind of program would redound to the benefit of coal mining in their home region while allowing them to build a national reputation as environmental advocates.

In fighting SMCRA, eastern firms also had to contend with the fact that the new regulatory regime was taking shape in the early 1970s. Environmentalists had set the national agenda and framed policy discourse. In this, they were abetted by key legislators and the news media, which played a crucial role in focusing national attention on the environmental problems associated with surface mining. Thus, the debate on SMCRA immediately focused on the question, what should a federal surface mining program look like? Despite very vigorous opposition early on, most large firms were drawn into this type of dialogue. To raise the question of the nation's need for SMCRA, eastern firms, especially by 1975 and the passage of S.425, would have had to derail years of legislative work and compromises. In a sense, small and medium-sized firms became outsiders in a policy debate that affected them severely.

The ideological climate tended to reinforce the "isolation" of eastern firms in the SMCRA debates. A possible way of undercutting the strength of pro-SMCRA forces might have been to raise the sensitive question of growth versus no-growth, thus painting SMCRA as a program that would hurt the majority of people. Indeed, eastern operators frequently argued that utility costs would skyrocket under SMCRA. However, the understanding among environmentalists and major business that this volatile issue would be kept out of public debates served to make such claims of eastern firms less credible. Also, eastern firms tended to hurt themselves in their public statements, sometimes revealing a political ideology completely out of step with the second half of the twentieth century. For example, B. V. Cooper, president of the Virginia Surface Mining and Reclamation Council and a director of NICOA, suggested to Morris Udall's Subcommittee on Energy and the Environment: "If the Congress is truly convinced that most of the citizens of the Nation want a Federal coal surface mining bill, then let us simply insert a provision which allows each state to decide for itself whether to participate in the Federal Program, such a decision to be the result of a popular vote during each State's next general election" (House Hearings, 1977b, v. 2, p. 360). One can only wonder at the thoughts of the congressmen as they listened to this theory reminiscent of John C. Calhoun's arguments on nullification and interposition.

The isolation of eastern firms was accentuated by the roles of trade associations. Eastern firms were concentrated in NICOA. NICOA's very existence indicates a different set of preferences from those firms with memberships in NCA and AMC. One independent operator, William Clements, argued explicitly that NCA did not represent him or his fellow independents. In congressional testimony (see Senate Hearings, 1973a, p. 708), he asserted that "the coal industry could survive the present attack of environmental laws if it could but speak for itself. How can the National Coal Association represent coal if its resources are derived from oil?" To substantiate this accusation, he pointed to the number of coal firms owned or controlled by oil companies. While this argument about oil's influence at NCA may be a bit overdrawn, it seems clear that NCA did not represent the interests of smaller eastern operators. This suspicion permeated SMCRA's legislative history. Norman Williams of OSM reported that a major problem in dealing with the coal industry was that small eastern firms saw a conspiracy among other larger and basically oil-related firms in the latter's "lukewarm opposition" to SMCRA (personal interview, July 28, 1980).

In addition, throughout the hearings and debates on SMCRA, eastern firms consistently appeared under the auspices of various state trade associations. Most prominent among these were the Mining and Reclamation Councils from Alabama, Kentucky, Tennessee and Virginia. This set them apart from other regions and served to portray eastern firms as selfish and parochial in their approach to SMCRA. An important undercurrent was the fact that many of the witnesses brought from eastern states to testify on behalf of SMCRA had been treated high-handedly in local politics by the state trade associations and their member firms. These coal firms had a strong enough political position at the state level to do this. However, at the federal level, as this behavior came to light, it made these firms appear irresponsible and cavalier about environmental issues. For example, Fred Kilgore of Virginia Citizens for Better Reclamation revealed that his organization had sent over four hundred letters to Virginia surface operators inviting them to a conference on reclamation and soliciting comments on a legislative proposal. He received one response (House Hearings, 1977b, v. 1, p. 6). This kind of situation, quite common in the East, cast those firms in a poor light before Congress.

The executive branch added little to eastern firms' prospects for success. None of the relevant federal agencies had anything to do with eastern operators. DOI focused on the West, the leasing and administration of public lands. Even the Bureau of Mines at DOI had no political or administrative relations with eastern firms outside of collecting data on coal production. Though EPA had some knowledge of eastern surface mining through its responsibilities for water pollution control, William Dickerson pointed out that coal mines were nonpoint sources and as such were not subject to direct EPA supervision (personal interview, July 21, 1980).

The one boost to eastern coal from the executive branch, of course, was the Ford vetoes in 1974 and 1975. However this victory was short-lived; President-elect Carter had pledged to sign SMCRA.

In sum, the high value eastern firms placed on defeating SMCRA had to be weighed against a low probability of success. Since their preference function boiled down to enactment or defeat of a federal surface mining program, the value of defeating SMCRA clearly seemed significant for eastern firms. On the other hand, they had to operate in a political-economic atmosphere clearly not conducive to their objective. With these countervailing pressures on their expected value of participation, the costs of participation became critical considerations.

The major point about cost is that many oppositional behaviors are quite inexpensive. Contacting a congressman or senator, writing letters, testifying before a committee or marshaling state and local support need not be prohibitive. Once SMCRA moved to the implementation phase, however, participatory costs escalated as the arena shifted to the courts and executive agencies. In these realms, participation required considerable outlays of time and financial resources.

At every hearing on SMCRA, eastern firms were represented by several witnesses. These included firms themselves, NICOA, state and local trade associations, state political leaders and bureaucrats and local congressmen. On occasion, NICOA arranged for local citizens or engineers from eastern states to present "expert" testimony. No other region consistently marshaled such a show in Congress.

Not surprisingly, the Virginia firms were the most visible and the most vocal throughout SMCRA's legislative history. Much of their participation was coordinated by the state trade association and the Congressman from southwest Virginia, William Wampler. Wampler consistently argued that SMCRA was intolerable, an economic disaster for Virginia. For example, he told a House committee: "This bill would have surface mining regulated in the same manner in our states with flat lands, in those with rolling hills, and in those mountainous regions of Central Appalachia. You must either admit that H.R.2 is not feasible, or it is an attempt to severely curtail surface coal production in the mountainous region of Appalachia" (House Hearings, 1977b, v. 3, p. 166). (Discussions with OSM officials indicate that there was an element of truth in his second conclusion.) In discussing firm preferences, we have seen already the tone and tenor of Virginia operators' testimony. Wampler tirelessly opposed SMCRA in every legislative forum. He cooperated with NICOA whose headquarters, interestingly, are in the coal region of southwest Virginia. In orchestrating the appearance of Virginia citizens to testify against SMCRA, for example, Wampler appeared with two coal miners from the UMW in District 28, which includes southwest Virginia. Both of these men presented short but dire accounts of what SMCRA would do to the economy of their home region.

Other Appalachian states also cooperated with NICOA and the Virginia firms in presenting strong and unified testimony against SMCRA. Representative Tom Bevill, appearing to testify with a group of Alabama surface operators and one miner from UMW District 20 in tow, told the Udall subcommittee: "I supported President Ford's vetoes of the bill passed in Congress in 1975. . . . We sometimes tend to forget sitting up here in Washington that there may be answers to problems that don't require our taking over an industry" (House Hearings, 1977b, v. 2, p. 70). Also from Alabama, William Kelce from the state's Surface Mining and Reclamation Council argued, "the enactment and continuing consideration of numerous environmental and safety laws and regulations . . . is severely eroding the economic and political stability of the United States mining industry" (House Hearings, 1977b, v. 2, p. 105).

The eastern firms also attempted to apply pressure on pro-SMCRA legislators they deemed vulnerable. In particular, they singled out Patsy Mink of Hawaii, chairwoman of the House Subcommittee on Minerals, Materials and Fuels. Two specific actions demonstrate both the depth of their hostility toward SMCRA and a certain lack of political sophistication. First, during the 93rd Congress, a group of Appalachian coal firms organized a demonstration against SMCRA in Washington. On the Capitol steps, they burned Patsy Mink in effigy, no doubt to the great amusement of larger coal firms and pro-SMCRA forces. Second, Virginia and Kentucky surface mining firms attempted to organize a boycott of pineapples to punish Representative Mink for her outspoken support of SMCRA. To their chagrin, they neglected to find out ahead of time that most pineapples were imported from Taiwan, not Hawaii.

In a similar vein, B. V. Cooper told Senator Henry Jackson:

> In 1975, just prior to the coal truck convoy which came to Washington protesting Federal surface mining legislation, well over 20,000 persons from Virginia's few coal producing counties signed a document opposing the bill. And you should know that the largest city in our coal fields has only about 4,000 people.
>
> I have furnished you today letters from Virginia's Lieutenant Governor, the Honorable John Dalton, and from the Honorable Andrew Miller who until recently was Virginia's Attorney General. Both letters urge that this legislation be defeated. Further, the Honorable Henry Howell, a former Lieutenant Governor, has stated his opposition to the bill (Senate Hearings, 1977c, p. 358).

This typifies the political behavior of eastern operators.

As SMCRA moved to its implementation phase, eastern operators continued their steadfast opposition, shifting their fight to the federal courts. The battleground they chose was the U.S. District Court in Richmond. NICOA supported the Virginia Mining and Reclamation Council in its suit against Secretary of Interior Cecil Andrus. Joined by several other state trade associations, they alleged that SMCRA was unconstitutional by virtue of the fact that it contravened the states' right to regulate land use in designating steep slopes as unsuitable for mining. Further, they charged that this amounted to the seizure of private property without due process or just compensation. Justice Glen Williams, in fact, ruled in favor of the Virginia operators. His decision was overturned in the circuit court of appeals. The Supreme Court, in July 1981, finally upheld the circuit court decision.

The courts may have felt comfortable with the pro-SMCRA argument explained by Dick Hall of OSM. Hall suggested that an imminent environmental threat can supersede states' rights or due process, and noncompliance on top of years of irresponsible commercial activity "somehow weaken the due process clause" (personal interview, July 29, 1980). The decision of the courts not only illustrates the lengths to which eastern firms would go in their opposition, it further illustrates their isolation within the political-economic framework. The legal action of eastern operators contrasts with the behavior of other types of firms that targeted specific provisions in SMCRA rather than trying to overturn the entire program.

At the same time, eastern operators continued to participate in legislative oversight activities. Under implementation, however, many West Virginia firms that had looked to SMCRA as a means of equalizing interstate competition joined in criticizing the program. Typically, James C. Justice, an independent operator from West Virginia, testified before an oversight hearing: "In my opinion, the members of the congressional committee that passed this legislation are guilty of gross misrepresentation to their colleagues in Congress . . . Congress should reappraise OSM's position as it is probably in violation of states' rights" (House Hearings, 1979, p. 122). Justice might have added that the Congress grossly misrepresented SMCRA to West Virginia surface miners who had thought the federal program simply would bring the rest of Appalachia up to Pennsylvania and West Virginia's standards. Frank R. Smith of the R. & S. Mining Company of West Virginia complained, for example:

> I am guilty of being too small. I cannot afford a lawyer or a $30,000–$40,000 man to sift through the regulations and help me comply. $12,000 in fines. The fight is about over for me. We all know the outcome. I cannot get a bond. I cannot get a permit. I have been told by consultants that a permit will cost me $50,000 to $100,000 and it will take 2 to 3 years to get it.
>
> The bank will not wait 2 to 3 years for me to make my payments (House Hearings, 1979, pp. 78–79).

The captive firms in the East, it should be noted, also became significantly more hostile to SMCRA in the implementation phase. The reason for this increased hostility lay in the amount of deep mining that most captives did. The emphasis on the surface effects of underground mining was significant under SMCRA, and it became clear that many captives' understanding that SMCRA did not target them was illusory. The vice president for corporate development of a utility company with nearly 100 percent underground operations noted that SMCRA's regulations on hydrological effects, road construction, spoil storage and refuse disposal came down just as hard on deep mines as surface mines. More important, deep mines entailed more surface construction and facilities, all of which OSM sought to regulate for surface effects broadly construed (personal interview, January 6, 1981). As all this became clear, captive firms and some medium-sized eastern operators expressed feelings ranging from disbelief to outright betrayal. One medium-sized operator suggested that OSM, with the aid and comfort of its congressional allies, attempted to renege on whatever compromises had been made in the legislative phase (personal interview, December 25, 1980). In fact,

Dick Hall of osm acknowledged that many field personnel in the agency exhibited "overzealousness or even stupidity" at the outset (personal interview, July 29, 1980). (This accords with the explanation Director Krevac gave of epa's early problems.)

Certainly, no firm had grounds for ignoring osm's duty to regulate surface effects of underground mining. However, the far-reaching and, in many cases, crass implementation of smcra hardened the position of medium-sized and captive firms in the East. As a consequence, the legal action taken by nca and amc and aimed at specific osm actions or proposed regulations received additional support. Captives also made their displeasure known at osm. Interestingly, those firms that had had dealings with epa (no small firms) even approached that agency to enlist its aid in "talking some sense into the new people at osm," this according to Director Krevac.

Judging from some of the behaviors of eastern operators, it might be tempting to dismiss these firms as either irrational or totally unsophisticated. Take the case of one operator who testified in 1971, 1973 and 1977. As late as 1977 he was lecturing a committee that had struggled with smcra for six years:

> Sgt. York, the most highly decorated G.I. in WW I, was from the mountain portion of a neighboring state, Tennessee. Upon his return to the United States, he found the entire Nation at his feet, offering untold riches. His only request of the Nation he had served so gallantly was *"Forty acres of bottom land."* This desire for level land is born in us hillbillies. . . . The provisions of this Act requiring that surface mined land be returned to its approximate original contour would deny us the opportunity to increase the amount of level land in our area (House Hearings, 1977, v. 1, p. 29).

This parochialism and shoddy reasoning (level land created from contour mining is not bottom land) might play well in the state capital of Kentucky. However, it was unlikely to sway Henry Jackson, Morris Udall or Patsy Mink. There is a difference between participating and participating effectively. Congressional testimony tends to be more persuasive when backed up by extensive scientific and technical information rather than folksy allusions to Sergeant York. Similarly, burning a chairperson in effigy or organizing misguided boycotts are unlikely to be effective.

On the other hand, eastern firms' ability to coordinate political activity, maintain a Washington office with Kilcullen as their hired gun and succeed at the state level of politics suggests that they were a politically aware and effective group. How can this portrait be reconciled with their, at times, amateurish political behavior?

First, it must be noted that much of what seemed amateurish was low-cost behavior. Organizing demonstrations or testifying before a committee a few times over a six-year period, especially considering the quality of their testimony, was relatively inexpensive. Therefore, despite a low probability of success, the combination of high outcome value and the low cost of oppositional behavior can explain some of this participation. Also, the cost of this participation was low because eastern operators had fought surface mine regulation in their home states before SMCRA came on the national agenda. Their arguments and political tactics had been relatively successful at the state level. Also, these fights had developed lines of communication and cooperation among the coal operators, especially in close-knit regions such as southwest Virginia. However, transferring their arguments to the federal level can appear amateurish. As we have suggested, the coal industry as a whole, much less eastern operators, had had very little to do with the federal government prior to SMCRA.

Second, and more important, we must draw a distinction between individual firm behavior and collective action among eastern firms. Much of the unsophisticated political behavior or weak testimony may be attributed to individual firms. They only had resources for low-cost behavior. However, the existence of NICOA and state trade associations could substantially reduce the costs of individual participation by providing economies of scale. This is especially important in understanding the costly legal actions undertaken in the implementation phase. Very few eastern firms could have justified those participatory costs.

The Midwestern Firms

Small and medium-sized firms of the Midwest did not have the same powerful incentives for participation and collective action that their eastern counterparts had. From the beginning, it was clear that the attention of environmentalists focused on surface mining in Appalachia and the West. None of the interviewees involved in developing SMCRA even mentioned the Midwest. Only the prime farmlands provisions dealt exclusively with the midwestern coal firms, and they proved more of a concern for large firms in the region. Moreover, pro-SMCRA forces showed a willingness to compromise on that issue rather than others bearing directly on the other two regions. Ostensibly, SMCRA, then, would not confront midwestern firms with the same dire consequences as their eastern competitors. Finally, the small and medium-sized producers in the Midwest did not enjoy the homogeneity of preferences found in the East. Many small firms in southern Ohio and southern Indiana were deeply concerned about the steep slope provisions. Many medium-sized firms in the central region were predominantly underground operations. And, more important, in the Midwest, there was no set of firms analogous to those of southwest Virginia that would spearhead collective action.

Like other large firms, those in the Midwest showed the greatest concern about those aspects of SMCRA that raised questions of risk and access. Their preference functions centered on public participation, responsibility for off-site hydrological effects and, in the particular case of the Midwest, prime farmlands provisions. The value of achieving some moderation of these provisions was high for midwestern firms. Lucrative contracts with public utilities for providing steam electric coal hinged on assured and expanding control of regional coal reserves. Large firms could afford a long-run view that incorporated the capitalized value of these reserves to be exploited in the future. The provisions cited above posed a threat to these future earnings by increasing the uncertainty about reserve control—public participation could tie up mining operations in legal action for years, and prime farmlands provisions could actually remove coal from commercial development. In addition, the lack of clarity about a firm's responsibility for off-site hydrological effects (i.e., nonpoint source pollution) could induce financial institutions to raise the price of borrowing to finance coal operations. If the immediate compliance costs did not alarm most smaller midwestern firms, neither should they have caused much concern among the larger firms. However, the aspects of SMCRA that increased risk and threatened access were quite another matter.

Turning to the probability of success midwestern firms faced, it was obviously higher than that of eastern firms. Reducing the impact of public participation, prime farmlands provisions or hydrological responsibilities was more within reach than overturning SMCRA. Also, the larger average size and higher productivity of midwestern firms allowed them to accept reductions in front-end costs or direct compliance costs as a significant political victory since neither threatened their viability. In short, the narrower aims of midwestern firms increased the likelihood of success. Not only were pro-SMCRA forces in lobbies and in Congress more likely to compromise on the concerns of midwestern firms without fear of gutting the program, but also large individual firms could afford experts who could carry on a challenge in sophisticated legal and technical-scientific terms.

The political and ideological climates surrounding SMCRA had little bearing on the probability of success for midwestern firms. Since they did not take an active role in SMCRA's development and since environmentalists concentrated on eastern and western surface mining, the political and ideological conflicts that surrounded eastern participation did not emerge.

Throughout SMCRA's legislative phase small, medium and large midwestern firms all kept a low profile. A major reason for this was the fact that environmentalists and their allies in Congress set an agenda focusing on issues relating to coal mining in the East and the West. Nevertheless, the behavior of other types of firms with more at stake or more resources than midwestern firms tended to obviate the need for their participation. Eastern firms, especially those from Virginia, Kentucky, Tennessee and Alabama, NICOA and NCA strongly opposed the provisions in SMCRA that would impose front-end costs on smaller midwestern firms. Similarly, large western and interregional firms took the lead in opposing citizens' suits, public participation and hydrology provisions, thereby relieving the demand on large midwestern firms to represent this point of view. Even in the case of prime farmlands provisions, almost uniquely midwestern in impact, midwestern-based interregionals acted vigorously.

Looking at congressional testimony during the legislative phase, we find no firms acting alone or in concert with other firms. John Kilcullen of NICOA did note that smaller firms from Indiana were members of the association and fully supported its efforts and those of eastern firms (personal interview, February 2, 1981). This is understandable given the topography of southern Indiana. However, none of the Indiana firms offered testimony. They simply paid dues to NICOA. As for larger firms, they supported the efforts of NCA and state trade associations, though again not with direct testimony (personal interviews, January 12, 1981 and January 2, 1981).

One action did distinguish some midwestern firms from others. At some point during SMCRA's legislative phase, many large midwestern firms developed staff positions to deal with environmental issues. Smaller midwestern firms did not. These organizational changes took the form of bureaus in engineering and public affairs departments or separate divisions of environmental affairs. If these divisions existed, their staffs were frequently augmented (personal interview, January 2, 1981). This is an important response to SMCRA because it equipped coal firms to fight environmental lobbies on their own terms. It also tended to freeze out smaller firms that could not afford to engage in the dialogue.

Once SMCRA moved into the implementation phase, the situation changed slightly. Though other firms could still carry the burden, midwestern firms played a more active role. When oversight hearings were held on SMCRA in 1978 and 1979, we see, for the first time, testimony from the Midwest in the persons of Representative Phillip Regula of Ohio and Paul Tostenson of the Ohio Mining and Reclamation Council. Congressman Regula's testimony exemplified the changed perception of many midwestern firms on SMCRA. His plea that OSM allow states with good reclamation programs (i.e., Ohio) to continue administering them without interference from Washington illustrated the rising concern in the Midwest over SMCRA's impact. Previously, Ohio firms and legislators had hoped that a federal reclamation program would weaken the competitive advantages of firms from Kentucky that operated in a more lax environment. They simply wanted Kentucky firms to play by Ohio rules. They did not envision a significantly more costly program for all firms. Tostenson's position, already cited in the discussion on preferences, also suggests great concern with the incremental costs under the regulation written by OSM. The Ohio Mining and Reclamation Council, like other trade associations in the region, had not anticipated the amount that a federal law would add in compliance costs.

The emphasis in this second phase on the surface effects of underground mining also increased the importance of amending SMCRA. In the estimation of one Illinois-based firm with roughly 40 percent surface operations, SMCRA added approximately $2.50 per ton to underground production costs. This was even higher than the $1.62 per ton reported for compliance with surface mining regulations (personal interview, January 2, 1981).

The extremely limited participation of midwestern firms is consistent with a rational calculus of participation. With much less at stake than other firms, rational midwestern firms opted not to participate. If smaller firms did participate, the basis of their behavioral decision would be the low cost of negative political action in the American political system and the fact that concern with specific provisions would be more likely to meet with political success than an attempt to overturn SMCRA. However, the provisions they would oppose, costly bonding and permitting procedures or prime farmlands provisions, were already opposed by other sets of firms: they could be "free riders" (Olson, 1965). The interests of western and interregional firms in the provisions of mutual concern to large midwestern firms was so great that the latter also could rationally behave as free riders.

This entire situation changed in the implementation phase. The much-higher-than-expected burdens combined with a reasonable probability of success in amending specific regulations and the low cost of opposition to increase the incentives for these firms to participate. It is in the implementation phase that midwestern firms and their state trade associations testified against OSM at oversight hearings. The only costly action undertaken by midwestern firms was a suit brought by the Indiana Mining and Reclamation Council. Similar to the Virginia suit, it challenged the constitutionality of SMCRA. However, we may dismiss this as atypical for the Midwest since southern Indiana has a great number of small surface operators working under similar conditions as their Appalachian neighbors. Indeed, there is little difference between surface mining in southern Indiana and in the neighboring states of Kentucky or Tennessee.

The Western Firms

SMCRA's provisions on citizen suits, public participation, alluvial valley floors and surface owner consent comprise the major elements of preference functions for western firms. Considering the scale of western coal surface mines, the investment sunk in them and the amount of financing at stake, any federal regulation posing a risk to ongoing or planned operations would jeopardize considerable assets. Removing any such risk necessarily would carry a high value. In the estimation of western firms the risks imposed by the above-cited provisions did pose a serious threat.

Given the political-economic framework of SMCRA's development, western firms could expect some success, if not total satisfaction, in mitigating the impacts of those provisions increasing risk or imperiling access to coal reserves. Also, western firms did not find themselves in the same adversarial position as eastern or even midwestern firms. First, the interest of western legislators in nurturing the western coal industry guaranteed western firms a sympathetic hearing in key committees of Congress. Second, the experience of western firms with tough state reclamation programs and with the federal bureaucracy through the leasing of public lands and the 211 regs made the firms and the environmentalists more responsive to one another. They spoke the same language.

They also shared a disdain for eastern firms. The environmentalists, of course, objected to the despoiling of Appalachia by eastern surface mining. Western firms resented the public perception of "the" coal industry that lumped them together with eastern operators. As an officer of one western firm noted, "Older firms [i.e., eastern, midwestern and interregional firms] couldn't adapt to the new climate, couldn't change their management philosophies" (personal interview, February 9, 1981). The president of another western firm complained that "NCA fought the same old battles and personified the enemy [to environmentalists]" (personal interview, February 5, 1981). This general view held by western firms and pro-SMCRA forces provided grounds for cooperation during the legislative phase.

The resources of individual western firms also proved to be an asset in regulatory politics. As a rule, western firms had departments specifically to deal with governmental and regulatory affairs. These departments provided western firms with environmental expertise and ongoing contact with Congress and the Washington bureaucracy, invaluable assets for participating effectively in the politics of environmental regulation. In addition, the economic and financial assets of these firms cast them in a positive light with pro-SMCRA forces. Large firms both could afford reclamation and fit into the perspective of the environmentalist movement that saw "big business" as a fact of life in the post–World War II political economy.

Finally, the leadership of western firms was an important resource in successfully influencing the outcome of SMCRA. In keeping with their different management philosophy from other sectors of the industry, some western chief executive officers could term themselves environmentalists, a label that ordinarily drew sarcastic remarks from eastern surface mining firms (personal interviews). One western president related that his firm cooperated with environmental groups in drafting and lobbying for a state reclamation program, adding that "it was better to be a part of regulation in the hope of getting a fair hearing" (personal interview, February 5, 1981). This outlook and behavior had produced reasonable state programs, and western firms apparently saw a good prospect for success at the federal level as well.

Western firms, indeed, anticipated some flexibility from pro-SMCRA forces in return for a cooperative legislative approach. The fact that many western firms sought amendments or alterations in very specific provisions that were not, for the most part, central to the program, only heightened that expectation. With a high value placed on mitigating potential regulatory effects and a reasonable probability of success, the cost of political action presented the only possible impediment to the participation of western firms.

During the legislative and implementation phases, limited individual behavior oriented to achieve very specific ends characterized the political participation of western firms. This reflected their preferences and their assessment of the likelihood of success in the political-economic environment of SMCRA's development. The preferences they valued highly directed them to seek concrete revisions in SMCRA rather than a wholesale redirection or total defeat of the program. Their estimation of the probability of success given the political-economic forces at work led them to put as much distance as possible between themselves and the rest of the coal industry. They attempted to do that in their public testimony and private dealings with legislators, congressional staffs and pro-SMCRA groups.

Of the nine western firms interviewed, six presented testimony before congressional subcommittees. Those six firms appeared at hearings independently, as opposed to eastern or interregional firms which usually appeared under the auspices of a trade association. None of the nine was a party to any lawsuits against SMCRA or OSM, and all indicated that it would be very unlikely for any western firm to become so involved.

An officer of one of these firms pointed out that to battle congressional advocates effectively in the legislative phase or OSM people on specific regulations required direct and personal contact (personal interview, February 9, 1981). According to Norman Williams of OSM, this particular firm understood how to accomplish things, a welcome change from the usual negative approach toward OSM taken by coal firms. By way of specific example, he noted that another western firm, Kemmerer Inc., operated a huge open pit mine in North Dakota and "quietly and reasonably negotiated a variance with us [OSM] on topsoil storage and regrading techniques" (personal interview, July 28, 1980).

Throughout the congressional considerations of SMCRA, western firms took a noticeably cooperative stance. The criticism they offered was, for the most part, specific and constructive. It also portrayed western coal mining as modern, reasonable and socially responsible, unlike other regional sectors of the industry. This positive tone reflected the general approach of western firms. They sought to get on with the mining of coal and felt that passing SMCRA in a reasonable form would clear up much of the uncertainty about coal surface mining's future. Dealing positively with pro-SMCRA forces, as one western firm official observed, depended almost entirely on a firm's reputation (personal interview, February 9, 1981). Recognizing this, he went on to note that western firms consciously stressed the fact that they started operations in the late 1960s under extremely stringent, by state standards, reclamation programs. Peter Gabauer of NCA noted, from his experience at the FTC and NCA, that, "As a general rule bigger, high-technology firms, those on the cutting edge of innovations, 'deal' with agencies—they play the regulatory game" (personal interview, March 25, 1980). This description clearly fit western firms.

Setting themselves apart from the rest of the coal industry, western firms put themselves in a strong position to argue for specific dispensations under SMCRA. Especially astute firms even sought advantages under SMCRA beyond their primary objectives of reducing risk and insuring access. W. P. Schmechel of Western Energy, for instance, not only detailed objections to citizen suits, alluvial valley floors and hydrological provisions. He also argued with regard to Title IV on Abandoned Mine Reclamation: "We believe this section should be modified to provide funding for reclamation of orphaned land from general revenues. This assessment of a reclamation fee of 35¢ per ton of coal produced by surface mining . . . is an unfair burden on coal mines in those western states where reclamation has always been an integral part of mining operations" (House Hearings, 1977b, v. 4, p. 147). Pursuing this same line of reasoning, Schmechel suggested: "If the Congress and the President are unwilling to provide funding for the reclamation of orphaned lands from the general revenues, then we ask that recognition be given to those states in the West for the reclamation plans they have developed and followed" (House Hearings, 1977, v. 4, p. 148).

These remarks reveal a number of interesting points characteristic of western firms' behavior in the legislative phase. First, they clearly illustrate an attempt to differentiate the mining and reclamation activity of firms operating in the West. Second, they imply that no matter what Congress decided on specific provisions, western firms stood willing and able to comply with almost any federal regulatory program. Third, these comments suggest how western firms sought to exploit their reputations as modern responsible surface miners at the expense of other sectors of the industry. The suggestion of using general revenues or, failing that, according some acknowledgment to western reclamation standards in assessing reclamation fees, demonstrates a constructive approach to resolving a problem. This is in stark contrast to the tactics of eastern firms, which demanded that states deal with reclaiming orphaned lands even though they had proven themselves inadequate to the task, or interregionals, which simply argued for a reduction in the 35¢ per ton fee.

Even more indicative of the extent to which western firms and pro-SMCRA forces could cooperate, Ira E. McKeever, the president of COLOWYO Coal Company, testified at the invitation of the Udall committee. In fact, his testimony was arranged by Donald Crane, a Udall staffer and subsequently western regional director of OSM. McKeever suggested that COLOWYO, a subsidiary of W. R. Grace and Company and Hanna Mining, had developed integrated mining and reclamation procedures that would comply with SMCRA. He noted: "In our judgment, we faced the likelihood of failure at great financial loss if we failed to identify, understand, and control acceptably the impacts of our operation on local community life and the environment" (House Hearings, 1977b, p. 200). He also testified:

> Our drilling and exploration has involved expenditures of over $1 million, and an additional planning cost total of more than $3 million (e.g., environmental impact statements). In addition, the purchase of land and capital equipment brings our total investment to $27 million. . . .
>
> As demonstrated, we mounted a massive effort to understand and control the effects of our mine operations. We sent copies of our study to a wide variety of public interest groups and met personally with their representatives to describe our plans and solicit their views (House Hearings, 1977b, v. 4, p. 202).

This type of outlook and testimony was unique to the behavior of western firms, though, as a rule, most of them did not carry cooperation and reputation-building to the lengths that COLOWYO did.

Upon SMCRA's enactment, most western firms expressed shock at the directions taken by OSM and those responsible for drafting particular regulations. They had assumed that, consistent with its language, SMCRA would place primary regulatory responsibility in the hands of state officials. In other words, they expected business as usual in the West. After the substantial delays in federal funding of OSM—the 95th Congress was tied up in a fight over the B-1 bomber—western firms felt "surprised at the specificity of engineering and design criteria in the regulations" (personal interview, February 9, 1981).

Surface mining firms found themselves in the awkward position of having to comply with performance standards as well as OSM's design criteria. If, after meeting an imposed design criterion, a firm could not meet a performance standard, it would still be liable under SMCRA. The president of a western firm felt that many of the compromises of the legislative phase were abrogated in implementation as "those who wrote the regulations went beyond the intent of the law; what they didn't get in the law, they got in the regulations" (personal interview, February 5, 1981).

The question is, what difference did this new attitude make in terms of western firm behavior? Even though, as a vice president of a western firm suggested, a dislike of OSM was the only unifying factor in the coal industry (personal interview, January 30, 1981), it was not a strong enough force to forge an alliance between western firms and other segments. Western firms engaged in no litigation against DOI and OSM. Rather they participated in regional or local hearings and presented their problems, complaints and comments to OSM's Washington office. There seemed to be a universal feeling that the local and regional officials of OSM had little or no understanding of mining operations: few had any engineering background. Western firms also complained about OSM's applying the letter of the law with no flexibility, citing as an example the shutting down of an entire operation for mislabeling topsoil storage areas. OSM officials in Washington corroborated this type of problem, though they felt it was not widespread (personal interviews, July 28, 1980 and July 29, 1980).

In light of these circumstances, western firms continued to rely on their reputations as responsible enterprises in appealing to Washington and maintaining close personal ties with OSM people. Given the "overzealousness" of some OSM personnel at first, it should be no surprise that western firms were dismayed about implementation. However, western firms restricted their behavior to commenting on regulations and cooperating with congressional oversight, while continuing to press for changes in the law or regulations that would reduce risk and increase access to coal reserves.

While western firms participated in the development of SMCRA, they did so on an individual and ad hoc basis. For these firms, the crucial factors in their calculus of participation were the importance of creating a more certain atmosphere for developing coal resources and the structure of the political-economic framework. The costs of participation presented no serious obstacle since western firms had developed environmental expertise, technical data and relations with state and federal regulators as a matter of course in surface mining coal. Trade associations or other forms of collective action would not provide significant economies of scale.

The value of resolving the controversy surrounding federal regulation of surface mining was critical for western firms with major investment projects in the planning or construction stages. T. L. Austin of Texas Utilities pointed out in hearings on H.R.2: "the conditions under which coal can be mined in the West have not been settled by the Congress and the President, so coal availability and cost . . . are indeterminate" (House Hearings, 1977b, v. 4, p. 262). Passing SMCRA would rectify this problem.

Given the nature of the political-economic framework, there were powerful incentives for western firms to pursue their objectives individually. First, the good relationship with pro-SMCRA forces and western legislators militated against collective action. Indeed, western firms consciously sought to dissociate themselves from the other segments of the coal industry. Their preferences clearly set them apart from those other segments. Even Ray Peck of NCA recognized this: "Western firms, mostly oil and utility companies or consortiums, are markedly different in their interests than firms in the Midwest or Appalachia" (personal interview, June 25, 1980). Therefore, it would be difficult for NCA to adequately represent western preferences. More important, NCA representation would have been a liability given the political-economic realities surrounding SMCRA.

Second, western firms had been successful in their dealings with environmentalists, regulators and legislators by operating through "quiet diplomacy" and supporting reasonable environmental regulation. Aligning with NCA or the other coal firms on SMCRA might preclude such political behavior. As in the case of eastern firms, asymmetries in the coal industry induced western firms to participate, but not through NCA. Like eastern operators, western firms displayed a unique and homogeneous set of preferences, though at the opposite pole from eastern operators. Western interests could be pursued best outside the mainstream of the coal industry.

The Interregional Firms

Interregional firms, both eastern-based and midwestern-based, faced the widest variety of consequences from SMCRA. This range of effects was reflected in the complexity of their preference functions. As large firms, however, both kinds of interregionals expressed a common concern over those impacts of SMCRA that created uncertainty or ambiguity and, thereby, jeopardized planned operations and financial arrangements. John H. Paul of AMAX summed up the overriding concerns of interregionals when he testified that SMCRA would "prohibit present or proposed mining, cause delay of mining operations, impose unnecessary burdens and costs, and allow for administrative interpretation of important sections which will result in more delays and continual litigation" (Senate Hearings, 1977c, p. 402). Although interregionals naturally resented regulatory costs they considered unwarranted and burdensome, their focus remained on the prospects for delay and restricted access to coal as well as uncertainty in SMCRA. At all hearings and in all public debates about SMCRA, interregionals consistently pointed to these factors.

In spite of their agreement on general objections to SMCRA, interregionals had varying specific interests. Depending on the location of their operations and the management philosophies of their leadership, interregional firms logically emphasized different provisions in SMCRA and took different political approaches.

If interregionals' preferences were complex and varied, the probability of success they confronted was not entirely clear either. Some factors in the political-economic framework operated in their favor. Others did not.

Of the factors which tended to increase their chances of success, the sheer size of interregionals loomed largest. First, as major national firms, they could mobilize considerable legal and scientific resources in making their case to Congress and later OSM. As Ray Peck of NCA pointed out, compared to smaller firms, major coal companies "had a more professional approach to legislative critique; they had legal and management resources" (personal interview, June 25, 1980). This benefitted interregionals in two ways. On the one hand, it cast them in a positive light relative to smaller firms. On the other hand, it allowed them to challenge pro-SMCRA forces with hard evidence developed by their own experts.

This kind of argumentation was persuasive, especially given the negative reference point of smaller eastern firms. For example, the two largest interregionals, Peabody and Consol, performed in-house analyses of SMCRA's cost impacts. Such studies were used effectively in commenting on specific provisions or proposed regulations, particularly in the later 1970s when the cost of regulation became a major concern of the Carter Administration.

Second, firm resources allowed interregionals to debate intelligently questions of revegetation, wildlife management or water pollution control. Most interregionals had engaged in some form of reclamation for decades. Some such as Peabody even had impressed OSM officials with their reclamation efforts. Environmentalists could feel "comfortable" with experts from larger firms. At a minimum, they spoke the same language, even if they held different ideological views. Moreover, the clear trend toward larger firms and industry concentration, along with environmentalists' views that major corporations represented the logical development of American business after World War II, tended to accord interregionals center stage throughout SMCRA's history.

Third, by virtue of their size, interregionals could make relatively forceful arguments about their contribution to the solution of the nation's energy problems and the tonnage of coal that SMCRA might cost. It is one thing for a firm producing 150,000 tons of coal per year from four mines to argue that it makes an important contribution to national security. It is quite another for a firm producing tens of millions of tons in six states to make the same argument.

Fourth, the size of interregionals provided them with a broader political experience than most smaller firms. They were not strangers to Washington or the federal bureaucracy. The acquisition of interregionals by petroleum companies reinforced this distinction. The oil industry had sustained close contact with the federal government throughout the twentieth century. Their political expertise and resources far exceeded that of even interregional coal firms. As this expertise and resource base became available to interregionals, differences between major and minor coal firms widened. In interviews, executives at interregionals regularly indicated that parent oil companies had offices in Washington and "kept an eye on the coal end of the business" (personal interview, January 2, 1981). The largest interregionals such as Peabody or Consol maintained their own Washington offices. In addition, all interregionals maintained close ties with NCA, whose board of directors included primarily executives from major firms.

Finally, the fact that almost all interregionals either operated or planned to operate western surface mines placed them in a favorable position to achieve their preferences that happened to coincide with those of large western firms. To the extent that environmentalists and key legislators were predisposed to be flexible toward those aspects of SMCRA bearing directly on western firms, the western operations of interregionals would benefit.

Despite their various advantages in the political-economic framework, the interregional firms' success was far from assured. Indeed, it is difficult to define success given their complex preferences. Their size and visibility was a two-edged sword. In addition to affording them political and technical resources, size and visibility cast interregionals as villains in the eyes of most participants. Interregionals had long dominated NCA, the most prominent trade association. Pro-SMCRA groups sought to structure SMCRA with avenues for ongoing public participation in order to counterbalance the political and economic advantages of interregionals. If smaller firms were the worst environmental offenders, pro-SMCRA forces feared that the danger from interregionals was that they, one day, might "capture" OSM. (The public attention to interregionals, incidentally, was precisely what western firms counted on in their strategy of dissociation.)

The importance of the ideological climate and the role of trade associations for interregionals changed over the course of SMCRA's history. This reflected, as much as anything, philosophical changes in the leadership of some major firms along the lines suggested earlier (pragmatic accommodationists replaced intransigent ideologues) and the reality that Congress eventually did enact SMCRA.

James Reilly, a vice president of Consol, told Senator Henry Jackson in 1968: "I am not ashamed of what happened with acid mine draining in the old mines of Pennsylvania. They made it possible for my daddy to have a three-room house and put me in it when I came to this country and to develop this Nation" (Senate Hearings, 1968, p. 134). Such a cavalier attitude about acid drainage from abandoned mines clearly was counterproductive in dealing with the 93rd Congress. The surprising support for abolitionist bills in the 92nd Congress made it imperative that interregionals rethink their ideological stance and political strategy of adamant opposition to SMCRA. Dan Deely, a former EPA official, attributed this "shift in the industry's attitude from frontier capitalism in 1970 to acceptance of the need for regulation" to "oil moving in without anti-government ideological baggage." Also, he noted, the development of western coal necessarily reoriented coal firms toward executive and administrative agencies from their traditional state legislative approach to regulation (personal interview, July 22, 1980).

A shift in NCA matched the emergence of a more politically sensitive approach among interregionals. Ray Peck, who served in the Commerce Department under Nixon and Ford, observed that NCA's capability in terms of staff size and expertise remained quite low during the formative years of SMCRA. In fact, he noted that until the Ford Administration's pocket veto of S.425, NCA had attempted no independent input into SMCRA. Its role was purely reactive and the Ford Administration developed its veto position with no assistance from NCA. As the 95th Congress was being seated, NCA augmented its staff substantially and helped organize an AMC-NCA joint committee to represent the coal industry (i.e., interregionals) in regulatory affairs. According to Peck, "The 1970s were a period of learning, reaction and development in the coal industry" (personal interview, May 12, 1980).

Both the interregionals and NCA owed a great deal to Gerald Ford who, through his two vetoes, stalled the implementation of SMCRA from 1975 to 1977. The White House played a critical role for interregionals in affording them the time to retool politically and reorient their approach. In one sense, this delay, especially the second veto that Congress fell three vetoes short of overriding, made pro-SMCRA forces less willing to compromise except with western firms that generally approved of SMCRA. One official at an eastern-based interregional went so far as to characterize the vetoes as a mistake since it would have been better to have Ford's people implement SMCRA than Carter's people, who had become exasperated with the vetoes (personal interview, October 30, 1980). This is the same firm cited earlier as having had a change in management philosophy. Its view showed a good deal of political acumen (20/20 hindsight?). Nonetheless, for NCA and most interregionals, the two-year hiatus allowed them to gear up for an effective role in the enactment and implementation of SMCRA.

For interregionals, a clear calculus of participation proved elusive. Certainly participatory costs posed no serious obstacle, especially after major petroleum firms with their vast political and financial resources entered the upper echelons of the coal industry. Also, the existence of NCA and the dominant position of interregionals on its governing board facilitated participation. NCA, as any trade association, has a vested interest in participation to justify its own existence.

The crucial problem in a calculus of participation for NCA and interregionals revolved around defining a preference function that would take account of SMCRA's crosscutting impacts. NCA also had to chart a behavioral course that would take advantage of favorable elements in the political-economic framework while minimizing the negative elements. There was no serious question of the value to interregionals of defeating or amending SMCRA. How to defeat or amend it and which aspects to focus on, however, proved to be knotty problems.

Because of these ambiguities, the behaviors as well as the preferences of interregionals remained ambiguous. Even the decision on whether to act individually or collectively could not be resolved easily. Different factors in the political-economic framework, individual preferences and firm leadership all pulled interregionals in different directions. Perhaps as a consequence of this ambiguity, the preferences as well as the behavior of interregionals, individually and collectively, shifted and even seemed somewhat contradictory over the course of SMCRA's history.

Undoubtedly, interregionals would have preferred to avoid all the manifold dilemmas imposed by a uniform federal surface mining act. The simplest expedient would have been to work diligently for the total defeat of SMCRA. This is exactly what they did between the Jackson hearings in 1968 and the defeat of the Hechler bill, H.R. 1000, in the 92nd Congress. In this enterprise, interregionals collectively opposed SMCRA, maintaining that state reclamation programs were adequate and that a federal program would be unmanageable. This negative and reactive behavior facilitated collective action among interregionals. It involved, almost exclusively, legislative testimony and minor lobbying efforts directed at individual legislators.

Their behavior conveyed the distinct impression that they did not take seriously the prospect of a federal surface mining program. Two factors underlay this impression. First, until the passage of NEPA and the creation of EPA, the political clout of the environmental movement remained suspect. Legislative activity seemed to be effective in derailing efforts to pass SMCRA as late as 1971 when the House and Senate considered a total of thirteen separate bills: H.R. 60; H.R. 444; H.R. 4556; H.R. 5689; H.R. 9736; H.R. 10758; S. 77; S. 630; S. 993 (the Nixon bill); S. 1160; S. 1240; S. 1498; S. 2455. Second, in 1968 and 1969, consideration and eventual enactment of the Coal Mine Safety and Health Act of 1969 (MSHA) occupied the attention of interregionals. MSHA was a major effort at federal intervention into the practice of coal mining and naturally diverted the attention of NCA and interregionals from SMCRA.

By the end of the 92nd Congress in 1972, it seemed clear that SMCRA would be enacted in the near future. It was only a matter of when and in what specific form. As the legal counsel for one interregional stated, "Our position evolved with SMCRA. In 1972, we became convinced that something would come out of Congress, though definitely not Hechler" (personal interview, October 30, 1980). The support received by Ken Hechler and his allies proved to be a watershed in the behavior of NCA and interregionals. This is not to suggest that by 1973 they reversed their position and endorsed SMCRA, or even that they adopted the stance of western firms. By 1973, however, a number of changes were set in motion. The new president of NCA, Carl Bagge, determined that NCA would beef up its organizational resources in order to "effectively confront the environmentalist threat" (personal interview, March 24, 1980).

Individual coal firms began a similar process of reorienting and augmenting their managerial staffs. Interregionals set about institutionalizing staff functions to specialize in political and/or environmental affairs. One interregional interviewed set up an Environmental Planning Office in 1972 to consult and coordinate on federal regulations (personal interview, October 30, 1980). Another established, in 1974, a Public Affairs Office that grew from fifteen to forty people in just two years. This same company increased its environmental engineering staff from twenty-five to two hundred employees between 1973 and 1975 (personal interview, January 5, 1981). A third firm created its Environmental Affairs Department in 1970, but its staff of four grew to sixty inside of three years, and the department was subdivided into air, water and noise divisions (personal interview, January 9, 1981).

No doubt the massive investment by interregionals in the West during the early seventies accounted for some of this activity. However, officials at a fourth interregional that had set up a corporate and regional environmental quality control center stated that "———'s structure, in part, reflects SMCRA's effect—increased attention to government affairs through the Environmental Department, legal staff, Government Affairs Office and engineering" (personal interview, January 12, 1981). This accorded with Dan Deely's observation that "coal companies started to take on action-oriented professionals," thereby facilitating a condominium with the environmentalists and a dialogue focused on problem-solving. These activities laid the groundwork for future participation in the SMCRA debates.

From 1973 on, argued Ray Peck of NCA, SMCRA "evoked a concerted industry-wide effort" (personal interview, May 12, 1980). This statement is at variance with our analysis that there were considerable differences in preference among coal firms. Even focusing solely on interregionals as Peck probably was, it is unclear that NCA reflected accurately a consensus position. A major interregional active in setting NCA policy "operated independently on SMCRA because NCA compromises . . ." (personal interview, January 9, 1981). Moreover, their increased internal resources in the early 1970s prepared interregionals for participation outside of NCA.

Nevertheless, throughout SMCRA's history, even after 1972, major interregionals appeared before Congress on panels under the auspices of NCA. How can this collective approach be reconciled with a reluctance to sacrifice for the greater good? One obvious answer is that collective action was more appearance than substance, perhaps calculated, as Norman Williams suggested, to retain a semblance of solidarity with smaller eastern and midwestern firms for UMW negotiations. Indeed, it is impossible to discount this factor. Yet, on careful analysis, a more subtle strategy appears to underlie collective participation.

Interregional coal firms participated both individually and collectively after 1972 because participation allowed them maximum flexibility in pursuing complex preferences, elements of which might even be competing rather than complementary. A veteran of the legislative battles over SMCRA, Dick Hall, suggested that NCA "led the bad guys to batter down the most objectionable provisions while preparing a more moderate group for compromise later on" (personal interview, July 29, 1980). The problem with this view is that the so-called bad guys helped set NCA policy and were, in fact, the firms that eventually compromised. It is important to recognize not only that NCA has its own institutional identity and interests coextensive with its permanent staff, but also that it cannot manipulate its major member firms. Those firms set policy on the board of directors. If Hall's view attributes too much power to NCA as an institution in its own right, it does raise the important point that NCA has an interest in appearing tough against environmental legislation. This certainly was President Bagge's perception, and it accorded well with the participatory needs of the interregionals.

An aggressive trade association afforded interregionals the luxury of taking the offensive on SMCRA while individually compromising on specific provisions or privately expressing qualified support for the program. Somewhat ironically, individual interregional firms could use NCA as a lightning rod for environmentalist criticism much as western firms sought to dissociate themselves from the rest of the industry. Also, whatever NCA achieved in "battering down the most objectionable provisions" would be all to the good. Interestingly, even if the chief executive officers of major firms acted as battering rams in public, this did not prevent lower echelon officials, especially those with technical expertise, from establishing a personal modus vivendi with the architects of SMCRA and the administrators of OSM. The same interregionals whose CEO lambasted the concept of federal regulation in public sent his technical staff to Washington to comment on and develop regulations to go with SMCRA.

Naturally, SMCRA's enactment caused the chief executives to moderate their rhetoric. But they continued to use NCA as a forum for collective opposition to SMCRA. In their collective participation, interregionals hammered away at provisions on public participation, citizen suits, steep slope bans, alluvial valley floors and prime farmlands. On the individual level, their participation was more pragmatic accommodation than implaccable opposition. As an official from an interregional noted, "——— derived the greatest benefit from working within the system, both independently and within NCA" (personal interview, January 12, 1980). Working within NCA, in addition to the arguments already developed, gave this firm legal standing for lawsuits as this official also pointed out.

Turning to the individual level of participation, how did this differ from collective behavior? Primarily, it differed in the specificity of firm interests and the arenas of participation. Individual action allowed interregionals to pursue interests not held in common with most other interregionals or to take initiatives without having to accommodate other firms. This individual behavior also reflected the increased political and technical resources of the firms. For example, one interregional recognized an opportunity to reduce its overall tax burden under SMCRA and seized upon it independently of NCA. Interregionals, as a group, steadfastly had argued for state-level regulation and, after 1972, for state primacy under SMCRA. However, this particular firm recognized that, on balance, given the geographical dispersion of its operations, it would benefit from a uniform 12.5 percent severance tax on coal tonnage and supported such an amendment. The amendment would have increased the tax on eastern operations, but this eastern-based interregional extracted significant tonnage from a mine in Montana where there was a 30 percent severance tax. By contradicting the NCA position and working for a uniform severance tax, this firm would derive an individual benefit (personal interview, October 30, 1980).

By the time most interregionals had developed the capacity and insight to act in the manner just described, SMCRA had been enacted. As a consequence, individual action focused on the implementation phase.

All of the interregionals interviewed revealed that the keys to individual participation were the development of ongoing personal contacts among pro-SMCRA forces, especially within OSM, and a regular liaison with Washington independent of NCA. Without exception, by 1977, the year of SMCRA's enactment, these firms had developed such capabilities. Each had formed a task force to assess SMCRA's likely impacts on firm operations and dispatched men from these task forces to comment on proposed regulations and work with OSM. As one firm discovered, its task force, in collaboration with OSM, had determined that "it could compromise on some issues" (personal interview, January 12, 1980).

All of the interregionals engaged in some form of litigation over SMCRA both individually and under NCA's umbrella. Nonetheless, people at OSM felt that large firms, on the whole, "developed qualified support for SMCRA." In Norman Williams' words, the interregionals "could afford to be more philosophical about SMCRA . . . lobbying and lawsuits came to target specific problems, leaving room to negotiate" (personal interview, July 28, 1980). To illustrate his point, he cited the fact that eastern-based interregionals had achieved some success in negotiating individual variances to allow mountaintop removal operations. Though Williams seemed most interested in conveying the notion that OSM was not an asylum for environmental lunatics, he indicated that a low-key well-documented case could carry weight, even from interregionals that had vigorously opposed SMCRA. Apparently interregionals recognized this. A midwestern-based interregional revealed that even before 1977 it had engaged in such contact with congressional staffers. The vice president for public affairs noted that "from 1974 ———— engaged in separate [from NCA] lobbying with a full-time staff in Washington, D.C." (personal interview, January 5, 1981).

Interregionals never expressed the support for federal regulation found among western firms. But as a practical matter, they ended up closer to western firms in their behavior than to eastern firms with whom they felt more of a kinship, due to their sharing a common history of commercial activity. This, no doubt, created a certain ambivalence among firm leaders. Dick Hall noted that "management at firms like Island Creek, Consol and AMAX remained . . . sharply opposed to SMCRA in principle, but they had to play ball or cash in their chips" (personal interview, July 29, 1980). Mixed metaphors aside, Hall put his finger on the fact that the new regime had dragged individual coal firms kicking and screaming into the arena of regulatory politics. Table 6.1 summarizes how firms acted during the development of SMCRA.

Table 6.1 Regulatory behavior by firm type

	Region			
Size	East	Midwest	West	Interregional
Large	Acted through NCA, maintaining opposition. Not a significant change from legislative to implementation phases.	Acted as free riders in legislative phase. Increased scientific/technical and legal staff in 1970s. Organizational adaptation. Increased individual opposition in implementation phase.	Cooperation with pro-SMCRA forces in hopes of flexibility at OSM. Seek specific amendments. Individual participation in hearings and administrative proceedings. Direct negotiation with OSM.	Evolving willingness to "play the regulatory game." Increases in scientific/technical and legal staff. Organizational and leadership changes to adapt. Collective opposition early on. Increasingly mixed strategy of collective opposition and individual accommodation. Seek specific amendments.
Medium	Low-cost opposition behaviors. Collective action under NICOA and state reclamation councils.	Acted as free riders except for southern Indiana and Ohio firms, which cooperated with NICOA in implementation phase.		
Small	Low-cost opposition behaviors. Collective action under NICOA and state reclamation councils.	Low-cost opposition behaviors. Collective action under NICOA and state reclamation councils.		

7   Social Outcomes

The enactment and implementation of SMCRA represented a major victory for environmentalists as well as a realization of the ideas and institutions of the new regulatory regime. SMCRA, for the first time, imposed a uniform set of performance standards and enforcement procedures on the coal mining industry. It went further than any new social regulation in establishing a framework for participatory democracy and machinery for public surveillance of business operations (Galloway and FitzGerald, 1981).

Because SMCRA's legislative history paralleled the emergence of the new regime, it provides a unique insight into regulatory change and shifts in government-business relations. Between 1968 and 1972, surface mining interests in general and coal firms in particular fended off attempts to pass a federal law. By 1972, however, the tide had changed. A new regulatory regime was emerging which proved more conducive to environmentalist goals. Advocates of regulatory change successfully interposed themselves into government-business relations under this new regime. They consciously sought to devise regulatory mechanisms and institutions that would reduce business' capacity to influence government. In its detailed design criteria for mines and surface structures, its provisions for public participation in permitting, its provisions on citizen suits, its exclusion from OSM of federal employees involved in the promotion or development of mining and its unusually high number of field enforcement personnel stationed in regional offices, SMCRA epitomized the regulatory ideal of the new regime.

Nevertheless, environmentalists were not completely victorious. Nor were coal firms completely vanquished. As we have seen, coal firms varied considerably in their preferences and behavior. So too did they differ in their views of the social outcomes of the fight to enact and implement SMCRA. Clearly some firms succeeded more than others in achieving their preferences. Variations in influence during SMCRA's development account for firms' different evaluations of the regulatory program.

Winners and Losers among Firms

A major concession won in the final version of SMCRA provided for grandfathering of financial commitments to construct surface mines. H.R.2 and S.7 both allowed a grandfathering clause for new operations or those already under construction. However, P.L. 95-87 exempted even those operations that were still in the planning stages but had received a formal commitment for financial backing. The commitment, moreover, could be from a parent company. While the exemption applied only to the Interim Program, it represented an important compromise.

Interestingly, Norman Williams revealed that "Morris Udall went to bat and won grandfathering for financial commitments as well as operations" (personal interview, July 28, 1980). On reflection, this makes perfect sense. Grandfathering financial commitments would benefit western firms or western operations of interregionals especially. Udall and other western legislators could, thereby, abet the development of coal mining in their constituencies without directly weakening SMCRA's environmental standards. It was primarily the huge western operations in which the threat of uncertainty had jeopardized multimillion-dollar financial arrangements. This grandfathering clause could allay the anxieties of financial backers and smooth the way for continued growth in western coal.

The provisions on alluvial valley floors were also eased. The original ban was dropped, although SMCRA left it to the discretion of the Secretary of Interior and the director of OSM to determine whether an alluvial valley floor was unsuitable for surface mining. Again, western firms were the obvious beneficiaries.

The ban on prime farmlands mining also was dropped in favor of the same discretionary structure. Although midwestern rather than western firms derived the benefits from this concession, the symmetry between prime farmlands and alluvial valley floor provisions made it difficult to relax one and not the other. This incentive to not "punish" the Midwest derived more force from the fact that clean air standards had hit midwestern coal mining the hardest of all three regions.

Coal firms also won concessions after SMCRA's enactment. For example, Dick Hall admitted that the regulation allowing firms ninety days to comply with an abatement notice was "too inflexible" (personal interview, July 29, 1980). Ninety days might be a reasonable time for compliance on a minor infraction. However, correcting nonpoint source pollution, for example, might require several months of testing and monitoring simply to trace the source, much less correct it.

OSM therefore decided to take a more flexible approach, not only on this specific regulation, but also with its general approach to enforcement and implementation. This decision in Washington, though, did not always bring about an immediate result among enforcement people in the field. The president of one western firm complained bitterly about the fact that he felt he had gotten satisfaction in Washington on some matters only to find a few months later that the situation at home had not changed (personal interview, January 30, 1981).

In the implementation phase, coal firms did achieve a major victory when an administrative law proceeding nullified forty-three regulations promulgated under SMCRA for the Permanent Program. This legal success enjoyed by NCA and several large firms contrasts with the defeat suffered by the Virginia and Indiana litigants. The major contrasts between these two legal challenges were the arenas chosen for court battles and the breadth of their allegations. The Virginia and Indiana operators viewed SMCRA as an unwarranted and unconstitutional intrusion of federal authority. Consequently, they challenged the legitimacy of the regulatory program itself. Larger firms, on the other hand, were committed, by 1977, to working with OSM on the regulations. They expressed no qualms about the constitutionality of SMCRA—at least not publicly. Their objections centered on the reasonableness of specific regulations or grants of authority to OSM.

Their legal challenge rested on detailed evidence developed by scientific and administrative law experts. They had the resources and capacity to mount such a challenge, whereas smaller firms did not. They also had a philosophical outlook more conducive to working with the federal authorities and within the system.

Reflecting on the successes, however limited, enjoyed by coal firms, a picture of the winners and losers emerges. The winners, if we can use that term for any firms subjected to such a massive regulatory program, were huge western firms. Interregionals too achieved some success, though much of it can be attributed to the coincidence of their preferences with those of western firms. Large and medium-sized midwestern firms achieved an important concession on prime farmlands, but this too was related to the influence of western firms via their success on alluvial valley floor provisions.

The losers clearly were smaller firms, especially those mining in Appalachia and using contour or auger techniques. None of SMCRA's provisions bearing on them was significantly relaxed. Front-end costs, incremental costs of compliance, the length of time required for planning and permitting and steep slope provisions all remained in full force by the final version of P.L. 95-87. To be completely accurate, SMCRA did provide for federal assistance to small operators in terms of meeting the data requirements for permitting. However, this did little to obviate the overall impact of SMCRA. This portrait of social outcomes accords with the assessment of Norman Williams. He saw western firms as the only real winners in the sense that "oil and coal interests in the West won out over farmer/rancher interests" (personal interview, July 30, 1980).

Firm Influence

Given these various outcomes from SMCRA's history, what can be said about firm influence under the new regime? Naturally, any such conclusions must be circumscribed by the fact that they are derived from a case study. More properly, they are only tentative conclusions. With that proviso in mind, a number of statements appear justified.

First, the winners under SMCRA exerted the most influence. Their preferences accorded most closely with the social outcomes. This situation suggests that under the new regime firm size correlated highly with firm influence. However this is almost universally acknowledged. The more intriguing proposition derived from the case study of SMCRA is that influence varies even among large firms. In fact, some western firms such as COLOWYO that achieved a good deal of success in the legislative and implementation phases seemed more influential than larger interregionals. Ostensibly, a certain size threshold is a necessary but not a sufficient condition for exerting influence under the new regime.

Beyond firm size two other factors distinguished firms' influence under SMCRA. The region of firm operations and firm leadership both help explain differential influence.

Region, in and of itself, cannot suffice as an explanation of firm influence. Nevertheless, the influential firms, the western firms, differed systematically from others. They uniformly employed higher capital-to-labor ratios in their operations, mined with the newest and most advanced technology and came into being relatively late, almost coincident with the emergence of the new regime. We should expect to find the same pattern of firm influence under any of the regulatory programs of the new regime. In addition, these factors should relate to influence across firms in different industries.

The important point is that the distinguishing characteristics of western firms means that they operate under different production and profit functions, which permit them to adjust more easily to regulatory costs and the political demands of the new regime. Under SMCRA, for example, western firms were more sanguine about federal regulation than other firms because their size and productivity gave them leeway to absorb additional costs. The more sanguine attitude left room for compromise. In a sense, they successfully wielded influence because their objectives were not so grandiose: they could live with SMCRA.

However, there was more to achieving their preferences than simply setting their sights low. The fact that western firms were new afforded them the opportunity to operate without the stigma of a negative public image. In the particular case of SMCRA, the newer firms developed in an atmosphere of federal regulation. They became attuned to administrative and regulatory process. They knew their way around the Washington bureaucracy as well as state legislatures.

This familiarity with regulatory affairs proved to be a great asset in communicating with environmentalists and legislators. Again, it contrasts with the approach of other firms. As the new regulatory regime emerged and interregionals became involved in western operations, they too developed such a sensitivity to regulatory processes.

Not all firms, however, could adapt. Others did not have the inclination to do so. In this regard, firm leadership comes into play. Firms that will go along to get along, firms that will take environmentalism and other issues raised under the new regime seriously will have a better chance of exerting influence on social outcomes. Western leadership recognized this early on, as we have seen. By 1977, many interregionals had undergone a change in management policy, if not philosophy, that put them more in the mainstream of regulatory politics under the new regime.

Others, however, did not change or did not go as far in their changes. They found it more difficult to influence regulatory policy. As administrators at OSM pointed out, a firm was more likely to acquire a variance if it calmly presented reasonable arguments that showed an appreciation of SMCRA's objectives (personal interviews, July 28, 1980; July 29, 1980; July 30, 1980). Such an approach to SMCRA depended on the outlook of firm management as much as anything else. As Dick Hall explained, "Coal was underregulated for a long time, and many firms came to view our regulations as a 'wish list' of the federal government. How can we be expected to give them the same hearing as a firm that makes serious efforts at reclamation and compliance?" (personal interview, July 28, 1980).

For the most part, Hall had in mind eastern firms operating in Virginia, Tennessee, Kentucky, parts of West Virginia and Alabama. Another official at OSM characterized these firms as having a "frontier capitalist mentality" (personal interview, July 28, 1980). The true flavor of this mentality emerged in an interview with the vice president of coal operations for a captive firm mining in Appalachia: "Some folks think we ought to put the land back the way it was for the Indians. Well . . . maybe they should go live in a wigwam on top of a mountain and pick berries. Personally, I like flush toilets" (personal interview, January 23, 1981). This bitter comment was made in response to the question: "Could you characterize your firm's relationship with OSM?" Such firms obviously lost in the legislative phase; and this strident attitude was unlikely to pay off with OSM in the implementation phase.

In sum, if SMCRA typifies regulation under the new regime, influential firms will be new, relatively capital-intensive, large and will view as legitimate new regulatory goals. The new kind of regulation, with its health and safety laws, environmental controls and consumer protection, requires significant capital expenditures for compliance. Just as important, it requires an increased political capacity to deal with regulatory agencies and Congress on issues outside the traditional concerns of private enterprise.

In a sense, the new regulatory regime met demands raised in the 1960s for "corporate social responsibility." These demands, consistent with the underlying ideology of the new regime, were directed toward a political economy in which government took a leading role in providing for the general welfare interpreted in the broadest terms. The new regime took for granted the political influence of huge firms in concentrated industries (coal was clearly moving in this direction) and sought to counterbalance it with regulatory controls to insure the public interest. Pro-SMCRA forces, like other public interest lobbies, had no faith in the market and, at best, ambivalent feelings about government's capacity to prop up the market. They felt comfortable dealing with monopolistic or oligopolistic enterprises because they viewed such firms as the natural development of a capitalist economy.

Indeed, there was a common interest binding together huge corporations and advocates of the new regulation: a desire for certainty. We have seen that certainty was a major concern for large firms confronted with SMCRA. It induced them to want some kind of law on the books so they could get on with mining coal. On the other hand, environmentalists had an interest in getting a law on the books to make certain that minimum performance standards would apply across the nation. This desire for certainty was not lost on Congress, which wanted to expand coal production. Morris Udall asserted: "I believe the Nation has got to increase the production of coal over the next decades. It is our insurance policy against the Arabs . . . but this uncertainty [about surface mining regulation] is paralyzing the country" (House Hearings, 1977b, v. 1, p. 49).

A similar community of interests applies to other regulatory programs in the new regime. Again, this favors the kind of large firms with the ability and management philosophy to interact with advocates of regulatory change. Such firms can be influential, and were in the development of SMCRA.

These social outcomes are depicted in table 7.1 below.

Political-Economic Changes under the New Regime

Addressing a conference in Washington, D.C., on coal mining and SMCRA, James R. Jones, the director of environmental quality at Peabody Coal, stated: "The regulations, which are our subject today, are just another in a series of actions over the past five to seven years which have changed the coal industry into one as regulated as is its principal customer, the electric utility. . . . But the industry *is* continuing to adjust to bring itself into compliance with these national requirements" (Jones, 1979b, p. 2). Jones' remarks indicated not only that the early 1970s witnessed the emergence of a new and wider kind of federal regulation of the coal industry, but also that the industry (read major interregional firms) was responding. This response of individual firms has added as much to the changing political-economic contour of the United States as the new regime itself. Both the emergence of the new regulatory regime and the firm response has brought about fundamental political-economic changes.

The initial changes were wrought by an interesting mix of traditional

Table 7.1   Regulatory outcomes by firm type

Size	Region			
	East	Midwest	West	Interregional
Large	Faced economic difficulty. Achieved allowances for mountaintop removal and head-of-hollow fill operations under certain circumstances. Could pick up market shares of smaller firms in the East.	Faced minor economic difficulty. Succeeded in eliminating prime farmlands ban. Failed to achieve significant reduction in public participation provisions. Successfully fought specific regulations.	Succeeded on the whole. Eliminated alluvial valley floor ban. Established credibility with Congress and OSM. Failed to reduce public participation. Success in court challenge to specific regulations.	Eastern and Midwest operations hurt, especially by regulation of surface impacts of underground mining. Shared outcomes of large firms in other regions.
Medium	Defeated in Congress and the courts. Faced likelihood of economic failures.	Defeated in Congress and the courts. Faced difficulty in adjusting to program requirements.		
Small	Defeated in Congress and the courts. Faced likelihood of economic failures.	Defeated in Congress and the courts. Faced likelihood of economic failures.		

legislative politics and new approaches. Public interest lobbies, citizen watchdog operations and environmental action groups injected a critical new element into legislative politics and policy making. Also, the emergence of the strong policy-making role for congressional staffs provided these new elements with access to more traditional channels of legislative politics. The alliance and overlapping membership between staffers and environmentalists proved crucial in getting the new regulatory program on the national agenda. However, it also subjected environmentalist goals to the political pressures and necessary compromises of traditional legislative politics. Thus, SMCRA came to focus on coal for political more than technical or legal reasons.

The fact that western legislators dominated the committees and subcommittees handling SMCRA meant that their constituencies' interests received close attention. In general, the legislative arena served to moderate environmentalist demands as they clashed with other legitimate interests. In a sense, the political system functioned as it was designed to, making momentous change difficult but not impossible (SMCRA eventually passed) while providing aggrieved parties several points of access to the policy-making process. The Ford vetoes and legal challenges as well as the anti-SMCRA role of several congressmen all illustrate the effective use of traditional political channels.

The response of firms to the new regime also has brought about fundamental changes in the political system. The firms that survived and coped with the new regulatory regime became politicized. As the federal government came to intervene in business firms' everyday affairs, these firms developed the capacity to deal with regulators and to have a more direct and early input into the development of regulation. This capacity is reflected in increased staffs, new internal structures and the investment of millions of dollars in public affairs. Individual firms also have assiduously developed a presence in Washington and direct personal contacts in the bureaucracy.

The new regulation has necessitated these changes. Trade associations can function effectively when there is a broad consensus, a homogeneity of preferences. The differential impact of the new regime tends to preclude this. Even if firms generally disapprove of a regulatory program, the nature of their individual economic/financial situation as well as the composition of the relevant subgovernmental framework may induce them to pursue their interests independently rather than collectively. Individual firms may see opportunities as well as penalties in a regulatory program. Commenting on SMCRA, an official of an interregional bluntly noted: "It's clear that the regulation is a boon to firms with the financial and managerial resources to comply. We can pick up the market shares of the smaller outfits" (personal interview, October 30, 1980).

In addition, individual participation makes it easier to wield influence since, under the new regime, trade associations tend to raise the suspicions and negative images associated with earlier regimes. By acting individually, firms need not overcome these stereotypes. They can portray themselves as fair-minded and responsible individually even if the trade association and other firms take a purely adversarial position.

Taking a more expansive view, individual firms themselves constitute new political actors. Now that firms have substantial individual political capacity it is reasonable to expect them to employ it. Based on the implementation of SMCRA, firms will attempt to maintain an ongoing presence in the new regulatory agencies. This presents a situation in which advocates of the new regime and individual firms will be extremely important participants in regulatory affairs.

8 Epilogue: The Deregulation of SMCRA

When President Jimmy Carter signed the Surface Mining Control and Reclamation Act in August 1977, it appeared possible to put aside a decade of legislative acrimony and implement a workable environmental policy. Larger firms had reached an apparent condominium with pro-SMCRA forces. This understanding, based on a common desire to get a statute on the books as well as some important concessions made by environmentalists, seemed to bestow a real sense of legitimacy on both SMCRA and OSM. Moreover, for the first time SMCRA had unqualified support from the White House.

The election of Ronald Reagan in 1980, however, radically altered the outlook for SMCRA, and for OSM in particular. Indeed, the Heritage Foundation, in a 1980 report on the Washington bureaucracy identified OSM as the leading candidate for deregulation, given the agency's record of "regulatory excesses" (*National Journal*, May 30, 1981, p. 17). President Reagan's appointee as Secretary of Interior, James Watt, cited this report approvingly in announcing his plans to revamp OSM and make it more "responsible." Equally ominous for OSM were the president's appointments at OSM directly. Walter Heine, the first director and an architect of the legislation, was replaced by James R. Harris. Harris, an Indiana state legislator, had participated in that state's constitutional challenge to SMCRA. The number two appointee, J. Steven Griles, brought a similar background to OSM. As a former official in Virginia's Department of Conservation and Economic Development, he represented perhaps the most ardent opponents of the federal surface mining law. The top two appointments, then, were drawn from the states stridently opposed to SMCRA.

From the standpoint of this study, the election of Ronald Reagan and the subsequent appointments of Watt, Harris and Griles to administer SMCRA present an interesting research opportunity. Under the new Administration, the institutional and procedural safeguards built into the statute would receive a stern test. In fact, SMCRA provides a formidable array of safeguard mechanisms under a comprehensive scheme of citizen participation. The Act itself specifically states that citizen participation would be critical to the enforcement of the law (Senate Report No. 128, 1977). We have seen already how extensive citizen involvement was written into the permitting process.

However, four of SMCRA's provisions deserve added emphasis. First, interested citizens not only may lodge complaints but also may participate in the inspection of mine sites. Second, citizens are assured access to all administrative processes since SMCRA extends the Supreme Court's criterion of "aesthetic values" (see *Sierra Club v. Morton*, 405 U.S. 727, 740, 1972) as a basis for citizen standing in state as well as federal proceedings. Third, SMCRA provides for the award of expenses incurred by citizens for participation in administrative proceedings (e.g., rule-making). These awards may be compensation for attorney or expert witness fees. Moreover, awards can be levied against a citizen and in favor of a coal firm only if the citizen acted "in bad faith." As L. Thomas Galloway, an attorney for the Natural Resources Defense Council, argues, "The one-way nature of the award of costs and expenses is designed to further the purposes of the Act, and to encourage citizens in their role as *private attorneys general*" (emphasis added, Galloway and FitzGerald, 1982, p. 262). Fourth and finally, citizens, with no restriction on standing, may petition OSM to initiate a rulemaking proceeding (30 U.S.C., sec. 1211 [q] 1979).

All of the safeguard mechanisms in SMCRA received a severe political trial with the election of Ronald Reagan. Just how well did these safeguards function in insulating SMCRA from a hostile administration?

What Went Wrong at OSM?

It is important to recall that adverse reaction to OSM was building even under the Carter Administration. Despite the propitious circumstances surrounding SMCRA's eventual enactment, criticism of OSM and its interpretation of the law had gained credance long before Ronald Reagan took office.

OSM's difficulties centered on coal firms' complaints that the agency, in writing the implementing regulations, went far beyond the spirit and the letter of the law. Many larger firms, even western operations tacitly or openly supporting SMCRA, became dismayed over both the regulations coming out of OSM and a perceived hostility at the agency toward surface coal mining. MARC also shifted its position from guarded support to open criticism once the regulations came on stream. It is important to keep in mind, though, that this criticism focused not on SMCRA per se, but on OSM, its reading of the law and its reputedly heavyhanded enforcement.

Under the auspices of Carter's Regulatory Analysis and Review Group (RARG), a task force was established to investigate the implementation of SMCRA. It reported that many OSM staffers and field inspectors overzealously pursued environmentalist goals, often using bureaucratic discretion to move beyond the intent of the law (New York Times, January 11, 1981, p. 1). Direction of the task force by Michael Petkas, a former associate of Ralph Nader, lent additional credence to the industry complaints. Apparently enthusiasm at OSM for regulating the coal industry struck even an architect of the new regulatory regime as excessive. This finding is entirely consistent with EPA Director Krevac's observation cited earlier, that excessive zeal is characteristic of new regulatory agencies.

OSM's zeal in fact was accentuated because it stood in contrast with the more moderate approach to environmental regulation evident at EPA. The contrast was brought into sharp relief over the issue of regulating nonpoint source water pollution. As we have seen, the EPA and Director Krevac steered clear of this problem both because of the scientific/technical difficulty of defining and regulating nonpoint source pollution and because of the political problems it could raise for the agency. OSM, on the other hand, took an aggressive approach. Though OSM could be lauded for tackling a serious and potentially hazardous environmental problem, it also left itself open to practical criticism and political attack since nonpoint source pollution was only a tangential and vague concern within SMCRA.

Nothing, though, reflected OSM's overzealousness more than its position on the question of design criteria for surface mining operations. Highly specific design criteria based on the BACT concept (best available control technology) clearly reflect the mistrust of market forces characteristic of the new regulatory regime. It also reflects environmentalists' profound suspicion of coal firms, particularly eastern operators.

As a matter of fact, the RARG report specifically recommended that

OSM rely on performance standards rather than design criteria in enforcing SMCRA. A further study of OSM, performed by the National Academy of Sciences (NAS) and commissioned by the Carter Administration, echoed the RARG conclusions. In addition to recommending that OSM adopt a performance standards approach, the NAS study suggested the use of market incentives to improve compliance and encourage innovation in reclamation technology (House Hearings, 1981a, pp. 162–163 and Senate Hearings, 1981b, p. 7.). Indeed the NAS report even suggested that under certain circumstances it might be acceptable to leave exposed highwalls in the western region. This proposal carried the notion of regulatory flexibility beyond the issue of design criteria to the concept of returning mined land to its approximate original contour, arguably the central goal of SMCRA.

OSM's interpretation of SMCRA, especially with respect to design criteria, also led it into a politically damaging confrontation with a number of governors. In particular, Governor Ed Herschler of Wyoming, a Democrat and a strong proponent of SMCRA, reacted bitterly to OSM's implementation of the law. His personal representative, Nancy Freudenthal, testified before the House Subcommittee on Energy and Environment:

> In the past, Wyoming has had serious disagreements with the Denver Office over OSM's proper role in life. We battled overzealous inspectors, letters to operators advising that spring had come for vegetation sampling, and day-to-day conflict over the appropriate permitting and regulatory requirements. I can assure you, Mr. Chairman, that Wyoming will not tolerate a repeat of these frustrations, and that State primacy will come either from the Secretary or from the courts (House Hearings, 1981a, p. 65).

The tone of this statement is as revealing as the substance. OSM had alienated governors in mining states, even governors who had supported SMCRA.

The major source of this controversy between OSM and the states was SMCRA's provision for state primacy, that is, the eventual transfer of implementation activities to the states. This primacy provision was an important compromise, specifically recognizing regional differences in mining and reclamation requirements. However, OSM's employment of the BACT concept as well as its insistence on highly specific rules for compliance effectively undermined the compromise.

OSM specifies that to achieve primacy states must develop programs "as stringent as" the federal law and regulations. This, in effect, made state primacy little more than window dressing. Since there is little incentive for states to adopt a program more stringent than OSM's given the prospect of economic competition from other coal-producing states, it was a foregone conclusion that state programs would be accorded primacy by duplicating OSM's. Under these circumstances, some states simply "ceded" authority to OSM's regional offices rather than undertake the development and implementation of their own plans. For once, Virginia reflected a consensus position on surface mining policy when that state's commissioner of mined land reclamation complained to a Senate subcommittee:

> Since the enactment of the law and the implementation of the interim regulatory program, the Commonwealth of Virginia has endeavored to institute a permanent regulatory program . . . in order to assume primacy. . . . Detailed requirements dealing with groundwater data collection, topsoil removal, haulroads, lands unsuitable for mining, sedimentation ponds, disposal of excess spoil and backfilling and grading . . . were dealt with through the so-called state window approach with the intent of providing alternative regulations which took the unique geological and topological conditions present in southwest Virginia into account. All of these alternatives . . . were disapproved by former Interior Secretary Andrus . . . as being in conflict with the provisions of the act (Senate Hearings, 1981b, p. 85).

While Virginia had opposed the law all along, almost all states ran into similar problems in dealing with OSM on the question of tailoring state programs to regional peculiarities. OSM felt that SMCRA's emphasis on uniformity logically dictated that the BACT concept be applied identically to all regions.

In a sense OSM's conflict with the governors and coal firms resembled the Soviet-American controversy over the Yalta Conference. Each side chose to emphasize different and inherently contradictory aspects of the agreement. In the final analysis, there may have been no more hope of reconciling OSM and its opponents on the issue of design criteria than reconciling the Soviets and the Americans on the issue of free elections in Poland.

The Reagan Challenge

If controversy began to swirl around the implementation of SMCRA prior to 1980, the election of Ronald Reagan directly challenged OSM and its implementation of the Act. Moreover, President Reagan's appointments of James Watt as Secretary of Interior and James Harris as director of OSM clearly signaled the new administration's captious view of SMCRA: having lost the court battle to declare the law unconstitutional, both appointees were intent upon radically altering the law's impact. Any alterations could be accomplished by either legislative reform or administrative action. Watt and Harris chose the latter.

The decision to deregulate by administrative action provided Watt and Harris with two important political advantages. First, the decision helped to foster a formidable coalition. Most of the criticism and the disillusionment about surface mining regulation developed around OSM and its interpretation of the Act. Therefore an administrative tack found allies for Watt and Harris among the governors and coal firms that perceived an arbitrary and capricious attitude at OSM. Second, eschewing the legislative alternative afforded Watt and Harris a good deal of flexibility and control. Rewriting those regulations that they deemed to be "particularly troublesome and overburdening without corresponding positive impact" (House Hearings, 1981a, p. 4) would avoid a protracted fight in Congress.

In particular, it would avoid a confrontation over changing SMCRA, a confrontation that would originate in the House Subcommittee on Energy and Environment and the Senate Subcommittee on Mines and Mining. While both of these committees would be hostile to amending the law, the House body would be especially so. Morris Udall and John Sieberling, arguably the architects of SMCRA, both sat on that subcommittee, Udall as chairman. Administrative action would be more expeditious and less openly defiant of key legislators. It also would afford Watt and Harris direct management of the deregulatory process.

Two distinct thrusts characterized the attempt to deregulate at OSM, reorganization and regulatory review. Secretary Watt and Director Harris introduced a sweeping reorganization plan that included a significant reduction in the number of federal inspectors. In addition, they initiated a thorough review of all regulations pertaining to SMCRA (30 CFR, chapter VII). In the course of this review they specifically solicited complaints from coal firms and trade associations about burdensome rules. This two-pronged assault on surface mining regulation was intended to bring substantial and immediate relief from the "troublesome and overburdening" regulations imposed by OSM.

Roughly outlined, the organizational structure confronting Watt and Harris when they took office was (1) a headquarters in Washington, D.C.; (2) five regional offices, each headed by a regional director, with one office in each of the nation's five coal producing regions; (3) forty-two field offices distributed throughout the five regions, roughly in proportion to the number of inspectable units in each region. The regional offices combined technical advisory personnel with administrative personnel. They also were responsible for helping states to comply with the interim and permanent regulatory programs, as well as assisting the states in drafting their own permanent programs. In addition, they dealt with mine operators, citizen groups and inspectors from the field offices. In a sense, each regional office operated as a semiautonomous mini-OSM.

Each regional office applied SMCRA in its own region. Indeed, precisely this situation angered many coal firms and governors. For their part, coal executives became frustrated at feeling they had reached understandings with OSM headquarters, only to have a regional director ignore that agreement. This was an especially sore point with western operators who had tacitly or openly supported SMCRA. Much of their ire was aimed at Don Crane, the regional director at the Denver office. Crane, who had worked on Morris Udall's staff in the enactment of SMCRA, was characterized by one western executive as "a czar unto himself" (personal interview, February 10, 1981).

Similarly, governors felt exasperated that a regional director could dictate policy to the chief executive officer of a state. Westerners again felt especially chagrined in this case because, although their states had tough surface mining laws before SMCRA, they had not achieved primacy after five years under the Act. Because of the key role of the regional offices in assisting states in the question for primacy, the Denver office came under heavy criticism. In a letter to Udall's subcommittee, Governor Ed Herschler of Wyoming expressed this view:

Wyoming had a good coal mine regulatory structure before the Federal Surface Mining Act was passed. . . .

You must understand that the State Program approval is critical. It recognizes that the State is the primary governmental entity for regulating surface coal mine reclamation in Wyoming. . . . If someone else has a different idea of state primacy than this, they will certainly have their share of headaches working in Wyoming (House Hearings, 1981a, pp. 117–118).

Against this background of dissatisfaction with regional offices, Secretary Watt's proposal for reorganizing OSM sought to close the regional offices and replace them with state offices. However, Watt admitted that this would be feasible only in regions where all states had approved regulatory programs (i.e., had achieved primacy). While Watt and Harris planned to approve state programs as expeditiously as possible, regional offices would remain open in four regions. This included the Denver office, which handled unresolved problems about federal and Indian lands.

The Watt plan, though, singled out the Denver office for special attention. The office was to relocate in Casper, Wyoming. Of all the aspects of the reorganization, this was one of the most controversial. It attracted a firestorm of criticism since it seemed like a punitive action directed against Don Crane and his staff. Chairman Udall expressed the concerns of many when he wondered: "There is so much work still to do that it is difficult for me to understand why anyone would want to throw the western region into chaos by putting it on the road from Denver to Casper" (House Hearings, 1981a, p. 7).

Despite Secretary Watt's contention that Casper is located near more coal (Wyoming has more coal reserves than Colorado), Denver is more centrally located in the region. Moreover, Denver is a federal regional headquarters with offices of BLM (Bureau of Land Management) and USGS (United States Geological Survey) with which OSM must work closely. Certainly the staff of the Denver office felt that the move to mobile homes in Casper from downtown offices in Denver was designed to lower morale and induce resignations. As one staffer at OSM's headquarters suggested: "Don Crane's people are just being punished for being tough. If Watt really wanted to be more efficient, he could cut staff more cheaply with RIF's [reductions-in-force] rather than sentencing them to Siberia" (personal interview, August 3, 1981).

Another important aspect of the proposed reorganization was the creation of Technical Service Centers in Casper and Pittsburgh. Previously each regional office had its own technical staff. These engineers and scientists were to provide assistance to states and coal firms, especially smaller firms. They also cooperated closely with the forty-two field offices in enforcing the regulations. Under the reorganization plan, all of the technical staff would be located in either Casper or Pittsburgh.

As Secretary Watt explained, his plan would significantly reduce the number of offices and the total staff of OSM: "Those 42 field offices will be cut to twenty-two offices with a work force of 650-plus people. In addition to that we will have two Technical Centers and we thought there was some logic to putting them where the coal was. That might not be logical, but we thought it was. So we picked Pennsylvania and Wyoming" (House Hearings, 1981a, p. 17).

Indeed, many felt the move was not only illogical, but also politically motivated. Congressman Sieberling, for example, pointed out: "I am told by people in OSM and in the States that the regional system allowed technical, legal and inspection staff to coordinate their work programs and carry out their functions. But under the new program, technical people would be isolated in two offices and the inspectors are going to be in State liaison offices" (House Hearings, 1981a, p. 26). While it is possible that the secretary's reorganization was designed to cut costs and facilitate state primacy, it seemed obvious that it also would create some confusion, induce experienced personnel to leave OSM, impede coordination among technical and enforcement staff and impede liaison between OSM and other DOI bureaus. These impacts could not fail to weaken surface mining regulation in the short run.

Secretary Watt and Director Harris sought to bolster the reorganization of OSM by rewriting the regulations governing SMCRA. The scope of this regulatory rewrite is indicated by the fact that in 1981, Harris' first full year, OSM initiated forty-two rulemaking proceedings touching on almost all sections of the surface mining regulations. There were, however, a number of key proposals made by the Reagan Administration that deserve special attention. These proposals were guided by three principles set forth by Harris at oversight hearings:

> Under my direction, the OSM's regulatory reform effort will acknowledge the broad goals of (1) increasing the States' role and responsibility in the regulation of surface mining and reclamation operations; (2) providing a framework for increased coal production to maintain the Nation's energy supply and; (3) performing the Agency's responsibility to provide a national framework for the protection of the environment from adverse effects of coal mining (House Hearings, 1981a, p. 8).

This statement clearly set forth not only the guiding principles of the Reagan Administration with respect to SMCRA, but also Watt's and Harris' priorities at OSM.

As indicated by Director Harris, the chief concern was to achieve state primacy as specified in SMCRA, section 101(f), which states: "the primary responsibility for developing, authorizing, issuing and enforcing regulations for surface mining and reclamation operations subject to the Act should rest with the States." This language tends to conflict with many of SMCRA's provisions which stress uniform national standards and oversight of state regulation. Nonetheless, 101 (f) clearly directs that state primacy should be pursued. The key regulatory change proposed to effect state primacy related to the so-called "state window" provisions, 30 CFR, 730.5 (b) and 30 CFR 731.13 (c) (2). Under the original language, state programs were to be "as stringent as" both SMCRA's provisions and OSM's regulations. The Watt/Harris proposal would change that language to "as effective as." The impact of this change would be twofold. First it would move OSM in the direction of adopting performance standards rather than design criteria as a regulatory benchmark. Second, and more important, the new language would facilitate the approval of state programs by providing more flexibility and acknowledging regional diversity in accordance with SMCRA, 101 (f).

In conjunction with the proposed state window change, Watt and Harris were determined to approve state programs as quickly as possible. To this end, they established January 1, 1981, as a deadline for approval of all state programs. This policy created a dilemma. On the one hand, the permanent regulations must serve as a standard against which the state programs are measured, irrespective of whether the criterion is stringency or effectiveness. On the other hand, Watt and Harris intended to approve programs while attempting to change the criterion. In order to overcome this problem, they proposed to accept state programs subject to subsequent review under revised regulations. This additional flexibility would follow from another specific regulatory change that would delete from 30 CFR, 503 (a) (7), the requirement that state regulations must be "consistent with the regulations issued by the Secretary (of Interior) under this Act (SMCRA)."

The second principle of the Reagan reform, increasing coal production, would be advanced by a number of specific regulatory changes. These changes all aimed, in one way or another, at opening up new lands to coal mining. In particular, Watt and Harris recommended regulatory changes with respect to prime farmland regulations, mining on federal lands and citizen participation.

Watt and Harris did not specifically recommend a rewrite of the prime farmlands provision in SMCRA. However, in his Senate confirmation hearings Director Harris opined that the 100 percent premining productivity standard set forth in SMCRA, sections 510 (d) (1) and 519 (c) (2), would be difficult to achieve. Rather he suggested that even 75 percent productivity on prime farmlands might be too high a target. This attitude raised serious questions about OSM's intent to enforce regulations prohibiting surface mining on prime farmlands unless the 100 percent standard could be assured. In fact, OSM cooperated with the state of Illinois in allowing surface mining on prime farmlands on an experimental basis, with no such assurance.

Coal mining on federal lands would be affected by three very important changes in OSM's regulations. First, Director Harris proposed a redefinition of "valid existing rights" to mine coal (VERS). He proposed, in fact, three alternative redefinitions, two of which would add considerably more federal land to that available for surface mining. Commenting on these proposed regulatory changes, the National Parks Service asserted:

> Section 522 (e) (1) specifically prohibits surface mining coal within units of the National Parks System, while section 522 (e) (3) further protects NPS units by prohibiting surface coal mining operations in proximity to public parks if such operations would adversely affect them. *Both of these sections are conditional on VER.* . . . Congress clearly did not want . . . surface coal mining operations taking place on lands having an overwhelming national interest (cited in House Hearings, 1982, pp. 614–615; emphasis added).

Section 522 of SMCRA moreover directed the Secretary of Interior to determine if lands adjacent to national parks would be unsuitable for surface mining. The two flexible redefinitions of VERs effectively would nullify this authority by placing all lands adjacent to national parks outside the reach of section 522 as long as deeds of ownership or leases could be produced by coal firms. By OSM's own admission, between 1.1 and 3 million acres of national forests and 100,000 acres of federal wildlife preserves would be opened up by these redefinitions.

The other specific regulatory reform proposals on federal lands also would open additional areas to surface mining. One of these would have interpreted the term "mining plan" as pertaining to the Mineral Leasing Act of 1920 (30 USC, 181) rather than SMCRA. The mining plan is the document submitted by coal firms for the permitting process on federal lands. The impact of the new interpretation would be twofold. It would authorize the USGS rather than OSM to review mining plans. It also would allow the Secretary of Interior to delegate review authority from the USGS to state agencies. Both of these results would be suspect in light of the legislative history of SMCRA. Section 201 of the Act specifically prohibits any federal agency whose mission is the promotion of mineral exploitation from playing a role in the implementation of SMCRA. The USGS clearly falls under this ban. Furthermore, section 523 reserves to the Secretary of Interior and prohibits from the states approval of mining plans on federal lands.

Finally, under the new regulations proposed by the Reagan Administration, OSM would expand dramatically its experimental practices program. Designed to encourage technological advances in surface mining and reclamation, the experimental practices program was intended for limited use. However, in letters to state agencies OSM sought to encourage the use of experimental practices by liberally interpreting the regulations. One such letter submitted as evidence at an oversight hearing stated: "Specifically, OSM is requesting the submission of experimental practices permits which place emphasis on technology which can be developed to permit alternative post mining uses in *steep slope areas*. Practices that include *highwall retention* may be considered" (House Hearings, 1982, p. 398; emphasis added). The significance of this letter lies in its reinterpretation of the experimental practices program to encourage surface mining in Appalachia and not to return land to its approximate original contour. Returning land to its original contour was, perhaps, the single most important provision in SMCRA.

The third approach to rewriting regulations that would have opened up more federal land focused on restricting citizen participation. Citizens were to play a prominent role in all aspects of regulation, permitting, inspection, declaration of lands unsuitable for mining and enforcement. Each of these roles, but especially declarations of unsuitability, could constrain surface mining operations. Secretary Watt and Director Harris moved to limit the chances of citizens and public lobbyists obstructing coal firms in their search for exploitable coal reserves.

For example, although SMCRA expressly limits mining permits to five years under section 506 (b), Watt and Harris proposed, in effect, to eliminate this requirement. The five-year limitation would be circumvented by allowing coal firms to get a five-year permit for a mine plan covering an area that could be mined for up to forty years. In this way a coal firm could simply renew the permit every five years rather than prepare a new permit each time. While this proposal would cut costs by eliminating numerous expensive environmental studies, it also shut out citizen participation at the appointed five-year intervals.

In another change, Watt and Harris sought to redefine section 522 (c) of the Act by limiting petitions for declarations of unsuitability to citizens with a "property interest" in the mining area. This obviously changed the meaning of the Act, which defined "interest" in the same terms as the Supreme Court decision in *Sierra Club v. Morton* (cited above). "Interest" like "standing" to sue included aesthetic as well as material values. To further limit citizens in pursuit of unsuitability declarations Watt and Harris also proposed to increase significantly the amount of information required to file a petition and to allow cross-examination of citizens by coal firm attorneys in unsuitability hearings. They also sought to shift the burden of proof from coal firms to citizens in such hearings.

The Limits of Regulatory Reform

As one might suspect, the ambitious plan to deregulate surface coal mining fell short of its full objectives. Watt and Harris did accomplish a lot—probably owing as much to their audacity as to any deregulatory mandate. Indeed, many environmentalists expressed amazement at how far and how fast the Reagan Administration proceeded in its attempt to ease the burden of environmental regulation. A typical complaint appeared in the newsletter *Environmental Action*: "Keeping track of and trying to fight all the attacks on the environment has left Washington environmentalists with little time for introspection. People have been more concerned with trying to minimize the damage than with wondering how we got into this situation and why not more of us anticipated just how bad things would be" (1982, p. 26). Despite the considerable successes enjoyed by the Reagan Administration, supporters of SMCRA could take solace in certain definite limits to deregulation. Two factors limited Watt and Harris in their attempts to deregulate. The first of these factors was the popular and professional environmentalist support for SMCRA. The second, and in many ways the most interesting, was opposition from coal firms.

A strong, if indirect, indicator of the support for SMCRA was the Reagan Administration's decision to pursue an administrative rather than a legislative reform strategy . The advantages of the administrative approach notwithstanding, it carries with it an important caveat, suggesting that it was a second-best strategy. Unlike legislative reform, administrative measures provide no guarantee of lasting deregulation. It is entirely possible that succeeding administrations could "ratchet up" surface mining regulation. In fact, Bardach and Kagan specifically argue that with respect to the new social regulation, ratcheting up is a far more likely outcome than ratcheting down (Bardach and Kagan, 1982).

This being the case, one would have anticipated a legislative effort from Watt, Harris and their supporters. Legislative action clearly was in the mind of the NCA's Stephen Young as indicated in his testimony at SMCRA oversight hearings in 1982: "specific recommendations with respect to legislation must necessarily wait until the conclusion of this current rulemaking effort. . . . When the time is ripe for legislative action, the Joint Committee [of NCA and AMC] would be pleased to help formulate the needed statutory changes (House Hearings, 1982, p. 481). Despite the successes enjoyed by the Reagan deregulatory effort, Carl Pope, director of political education for the Sierra Club, pointed out that the Administration was unable to accomplish what its "principal clients" (i.e., big business) wanted, namely statutory deregulation (cited in Robinson, 1982, p. 26).

While anti-SMCRA forces took Young's wait-and-see approach to administrative deregulation, one must wonder why legislative remedies, their apparent preference, were not pursued by Secretary Watt? One possible answer is that legislative reform would have been risky. The risk derived from the support for SMCRA not only among legislators and Washington lobbyists, but also among citizens of the coal mining states.

One of the striking things about all hearings surrounding SMCRA was the participation of state and local citizen's organizations concerned with surface mining. These groups regularly included Virginia Citizens for Better Reclamation; Citizens for the Preservation of Knox County (Illinois); Save Our Cumberland Mountains; Western Organization of Resources Councils; Council of the Southern Mountains; and the Appalachian Defense Fund among others. These organizations stood out not only because they represented grass roots participation, but also because of the regularity, intensity and preparedness of their testimony. Not surprisingly, most of the citizens groups were from Appalachia where coal firms had operated with notorious recklessness in and around private communities. After years of frustration with regulation at the state level, these citizens' groups were unwilling to countenance a weakening of the federal law. The depth and breadth of this grass roots opposition made legislative reform a gamble, the more so because environmentalists and their legislative allies were expecting an attack on programs such as SMCRA.

The public support for SMCRA was enhanced significantly by the activities and resources of national environmental lobby organizations. Acting as public watchdogs, these lobby groups provided an early warning system against deregulation of the legislative as well as the administrative variety. As an example of this role, the Natural Resources Defense Council, through a "friendly" attorney in OSM's Denver office, obtained and publicized a DOI memorandum directing OSM solicitors to "settle civil penalty collection cases without resort to litigation wherever possible." The memorandum also requested solicitors to identify enforcement actions at variance with Secretary Watt's policy preferences and, where such actions existed, "vacate the underlying enforcement action and consequent penalty" (House Hearings, 1981a, p. 412). Only a national environmental lobby, a permanent and professional organization with allies in OSM, could have exposed and ended this questionable practice.

Moreover, these national lobby organizations were equipped to provide legally and scientifically sophisticated responses to deregulatory arguments offered by the Reagan Administration. With respect to SMCRA, for example, the Environmental Policy Center had established a task force, the Citizens Coal Project. Staffed with full-time lawyers and environmental scientists, the Project could offer incisive analyses of deregulatory impacts.

In addition to a high level of expertise, public lobby organizations could rely on political experience and congressional connections in meeting deregulatory challenges. Many environmental lobbyists were ten-year veterans of the struggle to enact SMCRA. This fact lent them two distinct advantages. On one hand, the years of experience gave them an intimate knowledge of not only the Act's numerous provisions, but also the history and legislative intent regarding the provisions. Hope Babcock of the National Audubon Society reflects this kind of experience. After working for the enactment of SMCRA, she served as a deputy assistant secretary of DOI in charge of the Act's Federal Lands Program. She brought this backgound to the Audubon Society. On the other hand, their experience had established their credibility with important members of Congress. For example, Ed Grandis, director of the Citizens Coal Project until 1982, and L. Thomas Galloway of the Natural Resources Defense Council routinely appeared at oversight and budgetary hearings on SMCRA. It is clear from their repartee with the House and Senate committees that they enjoyed excellent relations with Senator Henry Jackson as well as Representatives Morris Udall and John Sieberling, all key legislators. Of course they and other environmental lobbyists had enemies as well as friends in Congress. However, in hearings even their enemies respected their expertise and experience.

Public lobbyists also had at their disposal a variety of institutional and legal provisions typical of the new regulatory regime and expressly designed to guard against deregulation by administrative action. We already have described the provisions for citizen participation in rulemaking, permitting, inspection and enforcement. Legal remedies and standing to sue under SMCRA offered further means of protecting the regulatory program. Each of these means of guarding against deregulation was a potent weapon in the hands of experienced environmental lobbyists.

The specificity of SMCRA made these prophylactic measures easier to apply. With performance standards such as return-to-approximate-original-contour and 100 percent premining productivity of prime farmlands, prohibitions on the delegation of authority to states and requirements for citizen participation all specified in detail in the Act, it was difficult to "rewrite" SMCRA by administrative action.

Environmentalists could compare specific deregulatory proposals to specific sections of the Act. This necessarily imposed limitations on what Secretary Watt and Director Harris could accomplish. Their attempt to delegate authority to the states for declaring lands unsuitable for surface mining was defeated by precisely this strategy. In twenty pages of detailed testimony, Hope Babcock of the National Audubon Society documented how the Watt/Harris proposals on the Federal Lands Program would violate specific sections of SMCRA (House Hearings, 1982, pp. 378–398).

Similarly, a close reading of the Act combined with regular testimony in oversight and full use of the provisions for citizen participation and legal action aided environmental lobbyists in defending federal regulations. They kept their channels for participation intact despite DOI efforts to reduce opportunities for unsuitability petitions and participation in rulemaking. They maintained the 100 percent productivity standard for reclaiming prime farmland. They maintained the "as stringent as" language as a standard for judging state programs against SMCRA, though Secretary Watt succeeded in adopting the "as effective as" language with respect to OSM's regulations. They also prevented the implementation of deregulatory proposals that would have exposed national historical sites to endangerment from surface mining. Through legal action, the National Wildlife Federation forced OSM to review for possible civil and criminal penalties more than eight thousand citations against coal firms. When OSM failed to comply, the Federation returned to U.S. District Court and succeeded in imposing a review schedule on OSM (*Business Week*, October 12, 1983, p. 40).

While the intense opposition to deregulation was to be expected from local citizen organizations and Washington-based environmentalists, it was surprising that opposition also arose from coal firms, even those firms that did not oppose SMCRA and had become outraged with the performance of OSM. However, the intuitively obvious expectation that coal firms would support deregulation is suspect for precisely the same reason as the expectation that coal firms would oppose SMCRA. The basis for firms' preferences and behaviors is their individual financial-economic situations. This pertains to deregulation as well as regulation.

Even so, one might argue that the election of Ronald Reagan so altered the political-economic framework of regulatory policy that we should expect a shift in firms' behaviors as their estimates of the probability of success in opposing SMCRA increased. Our model of firm behavior suggests that the Reagan challenge should affect regulatory policy in two ways, directly through changes in the political-economic framework and indirectly through encouragement of firms to believe that they could change federal surface mining policy.

This argument does hold for some firms. Those firms whose preferences reflected opposition to federal regulation, but whose political sensitivity had dissuaded them from aggressive behavior, logically should have become more militant as the Reagan Administration took over at OSM. In fact, coal firms in seven eastern states filed court injunctions against OSM within two weeks of the 1980 election. They sought to overturn OSM's rejection of state regulatory programs as deficient. Since these rejections had been issued almost two months prior to the 1980 election, these actions suggest a more militantly antiregulatory posture brought about by the Reagan challenge. However, significant opposition to deregulation persisted even among those coal firms whose preferences dictated support or accommodation.

A more circumspect view of deregulation developed, for example, among the larger firms, particularly those with operations in the West. Their opposition to the deregulation of surface mining fell into three distinct categories. First, they felt that a massive deregulation would promote uncertainty, a chief concern of theirs. Second, they feared a possible environmental backlash after the Reagan Administration left office. Third, they objected to the fact that firms that had not complied with SMCRA or OSM's regulations would be rewarded.

Undoubtedly, among coal firms the major reason for second thoughts about deregulation lay in the prospects for uncertainty that it raised. In 1982, Chairman Morris Udall opened oversight hearings before his Subcommittee on Energy and Environment by noting: "The Reagan Administration's decision to engage in a massive revision of the regulations governing the program has reopened old issues and raised new ones, and delayed the kind of regulatory certainty I had hoped would have been established by now" (House Hearings, 1982, p. 1). Major coal firms shared his appraisal and his dismay. According to a president of one large western firm, "Reagan and his crew launched the deregulation ship with high hopes, but for some of us on board it looks like we're in for a stormy voyage" (personal interview, January 7, 1983). Less cryptically, the vice president for governmental affairs at an interregional stated: "Sure we had gripes about OSM, especially Don Crane and his henchmen out in Denver. But the [Reagan] Administration is maybe throwing the baby out with the bathwater. We spent five years getting used to surface mining regulations—now everything is up for grabs. We like a lot of the changes, but it makes it difficult to plan major projects" (personal interview, January 17, 1983). Such comments echo those of larger firms on the enactment of SMCRA. These firms could live with and even prosper under SMCRA. However, they wanted regulatory certainty. Only in an environment of certainty could they undertake and carry through major investments.

The administrative strategy of regulation exacerbated the problem of uncertainty. Legislative reform, however risky, would have left the regulations as they were pending congressional action. More important, amending SMCRA to meet specific objections of coal firms would effectively foreclose the possibility of another major regulatory shift when a more environmentally oriented administration came to power. By simply reorganizing OSM and deregulating through rulemaking procedures, Secretary Watt and Director Harris left open the possibility of a subsequent ratcheting up of surface mining regulations.

The fear of an environmental backlash, in fact, prompted caution among many firms. At oversight hearings, Sheridan Glen, an assistant vice president of Arch Minerals Corporation and speaking for MARC, expressed this caution to Congressman Sieberling. In a moment of candor, Glen remarked:

I was listening to your remarks about the Act and the changes in the Act and it prompts me to comment . . . I really wish the industry had not fought the passage of this bill so hard . . . had we not fought so bitterly . . . in those years, I think we would be able to get the changes in the Law a lot easier now and we would be operating in a different regulatory climate with OSM so that we are not faced with the skepticism of you all in the environmental community that we are now (House Hearings, 1982, p. 81).

These comments are interesting in a number of respects. They come from a medium-sized firm that had fought bitterly against SMCRA. The fact that Arch Minerals appeared under the auspices of MARC rather than the more hard-line NCA with whom it was associated during the enactment struggles indicated its more enlightened approach. More important though, Glen revealed both a preference for amending the Act and regret at having fostered an environment in which amendment was so difficult.

Even if these comments were not completely ingenuous, it is safe to conclude that Glen and other coal executives wanted to put the acrimony of the 1970s behind them and to avoid a backlash in the future. Privately, some large firms were even more forthright in their fears of a backlash. The chief executive officer of a western firm frankly asserted: "Even if I opposed the Law—and I don't—I wouldn't want to whip up the environmental extremists. After the Reagan people leave town, who's going to prevent a vendetta against the surface miners? That's why we keep our noses clean and our distance from the shake-up at OSM" (personal interview, January 17, 1983).

It is important to note that none of the large firms interviewed objected to the reorganization and the reining in of the regional offices. Indeed they enthusiastically supported these measures. However, the massive review of the regulations and the more blatant attempt to reinterpret SMCRA caused them a good deal of anxiety about possible future reprisals. Consequently, they sought to dissociate themselves publicly from the deregulatory effort. Large firms, especially western ones, were conspicuously absent from SMCRA hearings after 1980; previously they had participated routinely.

Yet another source of coal firm dissatisfaction with the extensive proposals to deregulate surface mining stemmed from the fact that those firms that had complied with the Act and cooperated with OSM would be penalized, while those that had not would be rewarded. Recalcitrant firms, mostly smaller Appalachian and midwestern operators, would benefit from their obdurance by not having to meet the same stringent standards adhered to by their more responsible competitors.

These industry complaints centered on the way in which states administered SMCRA after having received primacy. The approval of permanent state programs under the state window concept was the most significant achievement of Secretary Watt and Director Harris. However, with minimal federal participation in the implementation of SMCRA, environmentalists' predictions of loose regulations and weak enforcement came to pass.

Stanbury, Reschenthaler, and Thompson have argued that the Reagan attempts to delegate authority over social regulation to the states leads to a "Gresham's Law of Regulation" (1982, p. 41). This argument applies particularly well with respect to SMCRA. As the states took over the program under the state primacy provision, regulation tended to fall to a lowest common denominator. This phenomenon led the director of governmental and environmental affairs at a large western firm to complain, "Some eastern states that never had a strip mine regulation to speak of before SMCRA tried to return to the good old days" (personal interview, February 3, 1984).

This situation created resentment not only across state boundaries, but also within states as lenient administration of SMCRA rewarded renegade firms. The problem and coal firm complaints came to a head in early 1984 when OSM was forced to take over inspection and enforcement operations in Oklahoma. Director Harris also threatened to do likewise in Tennessee and Kansas. Even the Reagan appointees apparently had to acknowledge the validity of firm complaints about state regulation of surface mining. Almost all firms preferred state to federal primacy in surface mining regulation. As the *Wall Street Journal* reported, though, "complaints about lax and unfair enforcement of mining regulations persists. Such complaints prompted a number of coal companies to take the extraordinary step of requesting temporary federal take-over of strip mining enforcement throughout Oklahoma" (February 14, 1984, p. 12). Clearly, OSM decided to make an example of Oklahoma because its coal industry was relatively small and was not crucial to the state economy. Nonetheless, the fact that Director Harris would retreat from his earlier unqualified support of the state window concept indicates the seriousness of the problem and the strength of feeling among aggrieved coal firms.

The Fruits of Deregulation

As with the enactment and early implementation of SMCRA, larger coal firms with partially or exclusively western operations fared well under the Reagan deregulation. The Reagan years did, however, differ from the Carter years in terms of how smaller firms and environmentalists fared. Unquestionably, small and medium-sized coal firms in the East and Midwest received at least short-run relief from surface mining regulation. Conversely, environmentalists suffered some setbacks, even though they managed to preserve many crucial regulatory measures.

Larger firms enjoyed nearly the best of all possible worlds under deregulation. They obtained a number of short-term benefits, without exposing themselves to the probability of reprisals from environmentalists in the future. Their behavior in response to the proposals offered by Secretary Watt and Director Harris again illustrates the advantages of individual firm participation in regulatory politics.

The reorganization of OSM reduced the intrusiveness of federal regulators in the day-to-day operations of surface mining. The reorganization also eliminated the semiautonomous behavior of regional offices— indeed it eliminated the regional offices. This reassured the larger firms that agreements worked out with OSM headquarters in Washington would stand up in the field. The granting of state primacy also served the immediate interests of the larger firms by placing regulatory responsibility in the hands of governmental officials interested as much in promoting coal production as in environmental protection. Unlike SMCRA, state laws did not strictly separate officials into reclamation and exploitation bureaus. In addition, state primacy encouraged competitive reduction of regulatory burdens among the states—a Gresham's Law of Regulation, indeed.

Finally, Secretary Watt and Director Harris did succeed in eliminating many design criteria regulations. In conjunction with this effort, the "as effective as" language adopted as a criterion in comparing state programs to OSM's regulations also advanced the concept of performance standards over design criteria. In general, larger firms, as a result of the deregulation, could operate in a more flexible environment and expect their petitions for variances or waivers to receive a more sympathetic hearing among federal as well as state regulators.

The advantages of the deregulation were not diminished by the prospect of an environmental backlash against these larger firms. By allowing the Reagan Administration, trade associations and more militant firms to push for a reduction in the regulatory burdens of SMCRA, cooperative firms could insulate themselves from the likelihood of retribution at the hands of succeeding administrations that might be receptive to environmentalist arguments. Most of the cooperative firms were interested in living under a workable regulatory program, not in altering SMCRA. To the extent that their behavior and comments demonstrated that interest, they succeeded in distancing themselves from those intent upon rewriting the law by administrative action. By keeping an arms-length relationship with the deregulatory efforts, individual firms could maintain their credibility with Congress as well as with OSM and environmentalists.

As for smaller coal firms, they derived the same regulatory relief as the larger firms, though for them it probably meant more. For example, the broadening of the experimental practices program to encourage surface mining on steep slopes allowed smaller firms to mine in places and with methods that had been denied to them under the Carter Administration. Because smaller firms tend to fill spot markets rather than supply coal under long-term contracts, the immediacy of regulatory relief was especially welcomed by them. Moreover, the approval of state regulatory programs allowed these firms to interact with state agencies that, prior to the enactment of SMCRA, had exhibited a paternal rather than an adversarial attitude toward the coal industry.

The impact of smaller firms dealing with more compliant regulators is illustrated well by the fact that OSM had to reassert enforcement authority in Oklahoma and threaten to do so in Tennessee and Kansas. In all three states small and medium-sized firms predominated. In all three states primacy led to environmental abuses and lax enforcement.

The most important conclusion to be drawn about the smaller firms, however, is that the fruits of deregulation will not necessarily last. Without statutory amendments, nothing prevents a more environmentally oriented administration in Washington from more strictly enforcing the law and regulations; or even from reasserting federal authority on a wide scale. To the extent that environmentalists and their allies in government believe that small coal operators are inherently irresponsible on environmental matters, it is reasonable to expect a different administration to deal harshly with those smaller firms that tried to ignore the law.

The gains of the coal firms notwithstanding, perhaps the best barometer of what the Reagan effort at deregulation accomplished at OSM is how the environmental interests fared in combatting it. At the outset it must be reemphasized that SMCRA remained intact. That, in itself, stands as testimony to the political ascendancy of environmentalism and the public lobby organizations. Secretary Watt and Director Harris did not effect a permanent deregulation of surface mining. What they did do was succeed in providing a strong measure of regulatory relief to coal firms by reorganizing OSM, approving state programs and emphasizing design criteria (all measures that, incidentally, the larger firms heartily endorsed).

Of these three accomplishments, only the reorganization stands out as a marked departure from the direction of OSM prior to 1980. Secretary of Interior Andrus and OSM Director Heine were approving state programs before 1980, though at a slower rate. The voiding of over fifty OSM regulations by a U.S. District Court suggested, moreover, that the agency ought to move away from design criteria and toward performance standards. Even before the Reagan appointees took over at OSM, reports by the National Academy of Sciences and the Carter task force on SMCRA had strongly recommended the same approach. In fact, the NAS report advised the adoption of economic incentives to promote responsible reclamation among coal firms. Thus, from the perspective of actual gains and losses, it seems that the environmentalists lost the battle but won the war of deregulation.

While not impregnable, the new regulatory regime appears to have resisted the attempt to deregulate surface mining. In this sense, the mechanisms for citizen participation and the public lobby organizations functioned as they were supposed to, to guard against an attack by a hostile administration in league with industry. Just as important, the specificity of SMCRA, characteristic of the new regulation, provided a bulwark against deregulation.

References

U.S. Government Documents

1968   *Surface Mining*. Hearings before the Senate Subcommittee on Minerals, Materials, and Fuels. 89-2. (Congressional Index Service.) Washington, D.C.: U.S. Government Printing Office.

1971   *Legislative Proposals Concerning Surface Mining of Coal*, prepared by the Congressional Research Service for the Senate Committee on the Interior and Internal Affairs, 92-1. (Congressional Index Service.) Washington, D.C.: U.S. Government Printing Office.

1972a   *Regulation of Strip Mining*. Hearings before the House Subcommittee on Mines and Mining in the 92-1. (Congressional Index Service.) Washington, D.C.: U.S. Government Printing Office.

1972b   *Surface Mining*. Hearing before the Senate Subcommittee on Minerals, Materials and Fuels in the 92-1. (Congressional Index Service.) Washington, D.C.: U.S. Government Printing Office.

1972c   *Issues Related to Surface Mining*, prepared by the Congressional Research Service for the Senate Committee on Interior and Insular Affairs, 92-1. (Congressional Index Service.) Washington, D.C.: U.S. Government Printing Office.

1972d   *Surface Mining Impacts*. Report prepared by the United States Geological Survey for the House Committee on Interior and Insular Affairs. 92-1. (Congressional Index Service.) Washington, D.C.: U.S. Government Printing Office.

1973a   *Coal Surface Mining and Reclamation*. Hearing before the Senate Subcommittee on Minerals, Materials and Fuels in the 93-1. (Congressional Index Service.) Washington, D.C.: U.S. Government Printing Office.

1973b   *Coal Surface Mining and Reclamation: An Environmental and Economic Assessment of Alternatives*. Report by the Council on Environmental Quality, Senate Document 93-8. (Congressional Index Service.) Washington, D.C.: U.S. Government Printing Office.

1973c   *Regulation of Surface Mining*. Joint Hearings before the House Subcommittee on Environment and Subcommittee on Mines and Mining in the 93-1. (Congressional Index Service.) Washington, D.C.: U.S. Government Printing Office.

1973d   *Regulation of Surface Mine Operations*. Hearings before the Senate Committee on

Interior and Internal Affairs, 93-1. (Congressional Index Service.) Washington, D.C.: U.S. Government Printing Office.

1973e   *Surface Mining Reclamation Act of 1973.* Senate Report 93-402 for the 93-1. (Congressional Index Service.) Washington, D.C.: U.S. Government Printing Office.

1975a   *Surface Mining Control and Reclamation Act of 1975.* House Report 94-45 from the Committee on Interior and Insular Affairs, 94-1. (Congressional Index Service.) Washington, D.C.: U.S. Government Printing Office.

1975b   *Surface Mining Briefing.* Hearing before the Senate Committee on Interior and Insular Affairs, 94-1. (Congressional Index Service.) Washington, D.C.: U.S. Government Printing Office.

1975c   *Surface Mining Control and Reclamation Act of 1975.* Senate Report 94-28 from the Committee on Interior and Insular Affairs, 94-1. (Congressional Index Service.) Washington, D.C.: U.S. Government Printing Office.

1975d   *Surface Mining Control and Reclamation Act of 1975.* Conference Report, Senate Report 94-101, 94-1. (Congressional Index Service.) Washington, D.C.: U.S. Government Printing Office.

1976   *Surface Mining Control and Reclamation Act of 1976.* House Report 94-1445, 94-2. (Congressional Index Service.) Washington, D.C.: U.S. Government Printing Office.

1977a   *Reclamation Practices and Environmental Problems of Surface Mining.* Hearings before the House Subcommittee on Energy and Environment, 95-1. (Congressional Index Service.) Washington, D.C.: U.S. Government Printing Office.

1977b   *Surface Mining Control and Reclamation Act of 1977.* Hearings before the House Subcommittee on Energy and Environment, 95-1, Parts 1–4. (Congressional Index Service.) Washington, D.C.: U.S. Government Printing Office.

1977c   *Surface Mining Control and Reclamation Act of 1977.* Hearings before the Senate Subcommittee on Public Lands and Resources, 95-1. (Congressional Index Service.) Washington, D.C.: U.S. Government Printing Office.

1978a   *Implementation of Surface Mining Control and Reclamation Act of 1977.* Hearings before the House Subcommittee on Energy and Environment, 95-2. (Congressional Index Service.) Washington, D.C.: U.S. Government Printing Office.

1978b   *Amending Surface Mining Control and Reclamation Act of 1977.* House Report 95-1143 from the Subcommittee on Energy and Environment, 95-2. (Congressional Index Service.) Washington, D.C.: U.S. Government Printing Office.

1978c   *Implementation of Public Law 95-87.* Hearings before the Senate Subcommittee on Public Lands and Resources, 95-2. (Congressional Index Service.) Washington, D.C.: U.S. Government Printing Office.

1978d   *Implementation of Surface Mining Control and Reclamation Act.* Oversight Hearings before the Senate Subcommittee on Public Lands and Resources, 95-2. (Congressional Index Service.) Washington, D.C.: U.S. Government Printing Office.

1978e   *Surface Mining on Federal Lands.* Staff report to the Senate Subcommittee on Minerals, Materials, and Fuels. 95-91. (Congressional Index Service.) Washington, D.C.: U.S. Government Printing Office.

1979   *Implementation of Surface Mining Control and Reclamation Act.* Oversight Hearings before the House Subcommittee on Energy and Environment, 96-1. (Congressional Index Service.) Washington, D.C.: U.S. Government Printing Office.

1980a *Implementation of the Surface Mining Control and Reclamation Act*. Oversight Hearings before the House Subcommittee on Energy and Environment, Committee Serial No. 96-43 (House Committee on Interior and Insular Affairs). Washington, D.C.: U.S. Government Printing Office.

1980b *Economic and Ecological Impacts of Federal Surface Mining Regulation*. Report prepared by ICF, Inc. for the House Subcommittee on Energy and Environment. 92-1. (Congressional Index Service.) Washington, D.C.: U.S. Government Printing Office.

1981a *Implementation of the Surface Mining Control and Reclamation Act*. Oversight Hearings before the House Subcommittee on Energy and Environment, Committee Serial No. 97-16 (House Committee on Interior and Insular Affairs). Washington, D.C.: U.S. Government Printing Office.

1981b *Implementation of the Surface Mining Control and Reclamation Act*. Oversight Hearings before the Senate Committee on Energy and Natural Resources, Committee Publication No. 97-77.

1982 *Office of Surface Mining Reclamation and Enforcement Budget Request for FY83*. Hearings before the House Subcommittee on Energy and Environment. 98-1. (Congressional Index Service.) Washington, D.C.: U.S. Government Printing Office.

Personal Interviews

Personal interviews were conducted between 1979 and 1981 with:

Peter Gabauer, Specialist in Regulatory Affairs, NCA
Joseph Lema, Transportation Expert, NCA
Raymond Peck, Chief Counsel for NCA
William Dickerson, Director of Office of Environmental Review, EPA
Joseph Krevac, Director of Water Criteria and Standards, EPA
Dan Deely, Nonpoint Source Pollution Program, EPA
Norman Williams, Assistant Director, OSM and Congressional Staff for Patsy Mink (D-Hawaii) and Lee Metcalf (D-Mont.)
Dick Hall, Director of Enforcement, OSM
John Kilcullen, General Counsel, NICOA
Dan Jerkins, Vice President, MARC

In addition, interviews were conducted at thirty-four coal firms of various sizes and types. Anonymity was a precondition to these interviews. Therefore, interviewees are identified in the text with respect to their position and the type of firm with which they are associated.

Works Cited

Ackerman, B., and Hassler, W. *Clean Coal/Dirty Air*. New Haven: Yale University Press, 1981.

Alchian, A., and Kessel, R. "Competition, Monopoly, and the Pursuit of Money." In Universities—National Bureau Committee for Economic Research, *Aspects of Labor Economics*. Princeton, N.J.: Princeton University Press, 1962.

Allaby, M. *The Eco-activists*. London: Charles Knight, 1971.

Anshen, M. *Corporate Strategies for Social Performance*. New York: Macmillan, 1980.

Averch, H., and Johnson, L. "Behavior of the Firm under Regulatory Restraint." *American Economic Review* 52 (December 1962): 1053–69.

Bagge, C. "The Changing Regulatory Scene." In D. Moynihan, ed., *Business and Society in Change*. Washington, D.C.: American Telephone & Telegraph, 1975.

Bande, E., ed. *The Corporation in a Democratic Society*. New York: H. W. Wilson, 1975.

Bardach, E., and Kagan, R. *Going by the Book: The Problem of Regulatory Unreasonableness*. Philadelphia: Temple University Press, 1982.

Barry, B. *Sociologists, Economists and Democracy*. Chicago: University of Chicago Press, 1978.

Bauer, R., et al. *American Business and Public Policy*. New York: Atherton Press, 1963.

Baumol, W. *Business Behavior, Value and Growth*. New York: Harcourt, Brace & World, 1967.

Beard, C. *Public Policy and the General Welfare*. New York: Farrar & Rinehart, 1941.

Becker, G. "Irrational Behavior and Economic Theory." *Journal of Political Economy* 70 (February 1962): 1–13.

Bernstein, M. *Regulating Business by Independent Commission*. Princeton, N.J.: Princeton University Press, 1955.

Berry, J. *Lobbying for the People*. Princeton, N.J.: Princeton University Press, 1977.

Bituminous and Lignite Coal Production (Minerals and Materials Monthly Survey). Washington, D.C.: U.S. Government Printing Office, 1970–80.

Blackman, J. *Business Problems of the Seventies*. New York: New York University Press, 1973.

Boulding, K. *The Organizational Revolution: A Study in the Ethics of Economic Organization*. Chicago: Quadrangle Books, 1968.

Brams, S. J. *Paradoxes in Politics*. New York: Free Press, 1976.

Business Week. Issues from 1970–83.

———. "The Shifting Contour of the Strip Mine Law," March 17, 1980.

Caldwell, L. *Man and the Environment: Policy and Administration*. New York: Harper & Row, 1975.

Caves, R., and Roberts, M., eds. *Regulating the Product: Quality and Variety*. Cambridge, Mass.: Ballinger, 1975.

Chandler, A. *The Visible Hand: The Managerial Revolution in American Business*. Cambridge, Mass.: Harvard University Press, 1977.

Christian Science Monitor. Issues from 1970–80.

Coal Age. Issues from 1970–80.

Coal Surface Mining and Reclamation: An Environmental and Economic Assessment of Alternatives—CEQ Pursuant to S. Res. 45 (A National Fuels and Energy Policy Study, 93-8). Washington, D.C.: U.S. Government Printing Office, 1973.

Commoner, B. *The Politics of Energy*. New York: Alfred A. Knopf, 1979.

Cost Analysis for Model Mines for Strip Mining of Coal in the United States (Staff—Bureau of Mines, US/DOI). Washington, D.C.: U.S. Government Printing Office, 1979.

Cost Impact Analysis of Select Provisions of the Office of Surface Mining's Permanent Regulatory Program. St. Louis: Consolidation Coal Company, 1979.

Craver, T. "On the Road to Regulation." *The Conference Board Record* 12 (October 1975): 20–26.

Cyert, R., and March, J. *A Behavioral Theory of the Firm*. Englewood Cliffs, N.J.: Prentice-Hall, 1963.

181

Dahl, R. A. *Polyarchy: Participation and Opposition*. New Haven: Yale University Press, 1971.
Danforth, J., and McAvoy, P. *Government Regulation: Where Do We Go from Here?* Washington, D.C.: American Enterprise Institute, 1977.
Davis, K., and Blomstrom, R., III, eds. *Business and Society: Environment and Responsibility*. New York: McGraw-Hill, 1975.
Department of Interior. *Diligent Development and Continued Operation: Federal Coal Leases* (Instructional Memorandum No. 80-274).
Derthick, M. *Policymaking for Social Security*. Washington, D.C.: The Brookings Institution, 1979.
Drucker, P. *The Concept of Corporation*. New York: John Day, 1972.
Edwards, F. "Managerial Objectives in Regulated Industries: Expense Preference Behavior in Banking." *Journal of Political Economy* 85 (February 1977): 156–62.
Elazar, D. *American Federalism: A View from the States*. New York: Crowell, 1966.
Energy and Environmental Analysis, Inc. *Benefit Cost Analysis of Laws and Regulations Affecting Coal* (prepared for Department of Interior). Arlington, Va.: Energy and Environment Analysis, Inc., 1977.
Environmental Action. Reagan & Company (July/August 1981).
Epstein, E. *The Corporation in American Politics*. Englewood Cliffs, N.J.: Prentice-Hall, 1969.
Financial Times. Issues from 1970–80.
Fiorina, M. *Congress: Keystone of the Washington Establishment*. New Haven: Yale University Press, 1977.
Forbes (magazine). Issues from 1970–80.
———. "All Stick and No Carrot" 119 (June 1, 1977): 72–74.
Franklin, B. "Stripping the West." *New York Times* (April 30, 1975): 14.
Galbraith, J. *The New Industrial State*. Boston: Houghton Mifflin, 1978.
Galloway, L., and FitzGerald, T. "The Surface Mining Control and Reclamation Act of 1977: The Citizen's Ace-in-the-Hole." *Northern Kentucky Law Review* (1981).
Gardner, J. *The Common Cause*. New York: W. W. Norton, 1972.
Garson, G. *Power and Politics in the United States*. Lexington, Mass,: D. C. Heath, 1977.
George, A. "Case Studies and Theory Development: The Method of Structured, Focused Comparison." In P. Lauren, ed., *Diplomacy: New Approaches in History, Theory and Policy*. New York: Free Press, 1979.
Goldstein, M., and Smith, R. "Land Reclamation Requirements and Their Estimated Effects on the Coal Industry." *Journal of Environmental and Economic Impact* 2 (1975): 135–49.
Gordon, R. *Coal in the U.S. Energy Market: History and Prospects*. Lexington, Mass.: D. C. Heath, 1978.
Green, M., and Moore, B. *The Closed Enterprise System*. New York: Grossman, 1972.
Greenwalt, C. "A Political Role for Business." *California Management Review* 2 (Fall 1959): 7–11.
Guzzardi, W. "What the Supreme Court Is Really Telling Business." *Fortune* 95 (January 1977): 146–54.
Haefele, E. "Shifts in Business-Government Interactions." A paper presented to the National Chamber of Commerce, Washington, D.C., 1978.
Harris, P. "The Socially Responsible Corporation and the Political Process." In M. Anshen, ed., *Managing the Socially Responsible Corporation*. New York: Macmillan, 1974.

Haveman, R. "Private Power and Federal Policy." In R. Haveman and R. Hamrin, eds., *The Political Economy of Federal Policy*. New York: Harper & Row, 1973.

Heclo, H. "Issue Networks and the Executive Establishment." In A. King, ed., *The New American Political System*. Washington, D.C.: American Enterprise Institute, 1978.

Henning, D. *Environmental Policy and Administration*. New York: American Elsevier, 1974.

Hirschman, A. *Exit, Voice and Loyalty: Responses to Declines in Firms, Organizations and States*. Cambridge, Mass.: Harvard University Press, 1970.

Hunt, L. "Regulating the Home Appliance Industry." In R. Caves and M. Roberts, eds., *Regulating the Product: Quality and Variety*. Cambridge, Mass.: Ballinger, 1975.

Jacobs, D., ed. *Regulating Business: The Search for the Optimum*. San Francisco: Institute for Contemporary Studies, 1978.

Jacoby, N., ed. *The Business-Government Relationship: A Reassessment*. Pacific Palisades, Calif.: Goodyear, 1975.

Johnson, J. *The Politics of Soft Coal: The Bituminous Industry from World War I through the New Deal*. Urbana: University of Illinois Press, 1979.

Jones, J. "The Process of Developing a Western Coal Mine." A paper presented to the National Western Mining Conference and Exhibition of the Colorado Mining Association, Denver, 1977.

———. "Meeting the Requirements of the Surface Mining Control and Reclamation Act." A paper presented at the Academy of Industrial Mining Engineers, New Orleans, February 18–22, 1979 (a).

———. "Surface Mining Regulations: Impacts." A paper presented at the Outlook for Coal—Promises, Promises, Energy Bureau Inc., Washington, D.C., April 30–May 1, 1979 (b).

Kahn, A. *The Economics of Regulation*. New York: John Wiley & Sons, 1970.

Kaysen, C. "The Corporation: How Much Power? What Scope?" In E. Mason, ed., *The Corporation in Modern Society*. New York: Atheneum, 1969.

Kemp, K. "Industrial Structure, Party Competition and the Sources of Regulation." In T. Ferguson and J. Rogers, eds., *The Political Economy: Readings in the Politics and Economics of Public Policy*. Armonk, N.Y.: M. E. Sharpe, 1984.

Kohlmeir, L., Jr. *The Regulators*. New York: Harper & Row, 1969.

Kolko, G. *Railroads and Regulation: 1877–1916*. New York: W. W. Norton, 1965.

Kristol, I. *Two Cheers for Capitalism*. New York: Basic Books, 1978.

Krutilla, J., and Fisher, A. *Economic and Fiscal Impacts of Coal Development: Northern Great Plains*. Baltimore: Johns Hopkins University Press, 1978.

Lakoff, S., and Rich, D., eds. *Private Government*. Glenview, Ill.: Scott, Foresman, 1973.

Latham, E. *The Group Basis of Politics*. New York: Octagon Books, 1965.

Leone, R. "The Real Costs of Regulation." *Harvard Business Review* 55 (November/December 1979).

———, and Jackson, J. "The Political Economy of Federal Regulatory Policy." Unpublished research paper.

Lilley, W., and Miller, J. "The New Social Regulation." *The Public Interest* 47 (1977): 49–62.

Lindblom, C. *Politics and Markets*. New York: Basic Books, 1977.

Lowi, T. *The End of Liberalism: The Second Republic of the United States*. 2nd edition. New York: W. W. Norton, 1979.